Gunnar Nützmann

Paolo Viotti

Per Aagaard

# Reactive Transport in Soil and Groundwater

## Processes and Models

Gunnar Nützmann
Paolo Viotti
Per Aagaard
(Editors)

# Reactive Transport in Soil and Groundwater

## Processes and Models

With 113 Figures

Springer

**Professor Dr. Gunnar Nützmann**
Institute of Freshwater Ecology and Inland Fisheries, Dept. of Ecohydrology
Müggelseedamm 310, 12587 Berlin, Germany
E-mail: *nuetzmann@igb-berlin.de*

**Professor Dr. Paolo Viotti**
University of Rome, Dept. of Hydraulics, Transportation and Roads
Via Eudossiana 18, 00184 Rome, Italy
E-mail: *paolo.viotti@uniroma1.it*

**Professor Dr. Per Aagaard**
University of Oslo, Dept. of Geosciences
Sem Sælands vei 1, 0371 Oslo, Norway
E-mail: *per.aagaard@geo.uio.no*

Library of Congress Control Number: 2005928442

ISBN-10 3-540-26744-1 **Springer Berlin Heidelberg New York**
ISBN-13 978-3-540-26744-7 **Springer Berlin Heidelberg New York**



**Springer is a part of Springer Science+Business Media**
springeronline.com

Printed in The Netherlands



Cover design: Erich Kirchner
Production: Luisa Tonarelli
Typesetting: Camera-ready by the editors

Printed on acid-free paper 30/2132/LT – 5 4 3 2 1 0

# Preface

Founded in 1971, COST is an intergovernmental framework for European cooperation in the field of scientific and technical research, allowing the co-ordination of nationally funded research on a European level. COST actions cover basic and pre-competitive research as well as activities of public utility. In particular COST action 629 is focussed to improve the scientific base for the development of integrated indicators of the environmental risks created by presence of pollutants in water with emphasis on the water body of natural porous media. Establishment of a set of integrated indicators to evaluate the pollution status and risk of the European water resources will aid environmental agencies, administration and regulators considerably and profit the society as a whole.

In May 2004 an International Workshop subjected to 'Saturated and unsaturated zone: integration of process knowledge into effective models' was organized at University of Rome "La Sapienza", Italy. It results from the joint activity of working groups 3 and 4 of COST Action 629 but have contributions from all other working groups as well. Working group 3 "Biogeochemical dynamics from soil to groundwater" concentrates on the study of processes that influence sorption, mobility and persistence/degradation of pollutants in groundwater and thus the input pathways from soil to groundwater. Working group 4 "Modeling of reactive transport in soil and subsoil" including both the transport and the soil interaction phenomena.

The need for a profound discussion on coupling processes and modeling stems from the major problems related to contamination of soil and subsoil in the EC countries and elsewhere. Technologies for remediation and restoration require deep knowledge and understanding of the involved processes with an adequate modeling approach in order to reach the plausible results which are applicable in practice. These topics are central to COST 629 objectives and the focus of this workshop was to increase our integrated process understanding and modeling capabilities and the key questions are how "processes" can be integrated into "modeling" and how to integrate processes when working at different scales.

Topics for the conference included:

- Numerical methodologies for the simulation of reactive transport from soil to groundwater

- Procedures for deriving model parameters from lab or field studies
- Procedures and protocols for model calibration
- Examples of coupled models (soil, subsoil, groundwater) for describing reactive transport
- Integration of process studies and numerical modeling
- Influence of scales on biogeochemical processes
- Understanding biogeochemical processes with modeling.

Though the conference covered the entire field of the topics which were mentioned before, three special focuses were selected, namely the process knowledge from laboratory and field studies, the modeling method, and the connection of the one to the other one. The relevant papers have been included into these three chapters, named *processes*, *modeling*, and *integration*.

This book contains 20 papers selected out of 67 pre-selected papers from both oral and poster presentation. The selection was performed by an Editorial committee consisting of members from working groups 3, 4 and COST 629 management board, and a subsequent peer-review process of each paper. The Editors would herewith like to thank the Editorial committee and the reviewers for their effort. We also thank Barbara Kobisch and Luisa Tonarelli for their technical help and proper relief during processing this book.

The Editors:

**G. Nützmann**

*Institute of Freshwater Ecology and Inland Fisheries, Department of Eco-Hydrology and Geographical Institute, Humboldt University of Berlin, Germany*

**P. Viotti**

*University of Rome 'La Sapienza', Department of Hydraulics, Transportation and Roads, Rome, Italy*

**Per Aagaard**

*University of Oslo, Department of Geosciences, Oslo, Norway*

# Contents

# List of Contributors

**Aagaard, P.**
Department of Geosciences, University of Oslo, P.O. Box 1047 Blindern, N - 0316 Oslo, Norway

**Aravena, R.**
Department of Earth Sciences, University of Waterloo, 200 University Avenue West, Waterloo, Ontario, Canada

**Aulenta, F.**
Dept. of Chemistry, University "La Sapienza", P.le Aldo Moro 5, I - 00185 Rome, Italy

**Besse, P.**
Dept. of Chemistry, Laboratoire Synthèse et Etude de Systèmes à Intérêt Biologique, UMR 6504 CNRS, University Blaise Pascal, 63117 Aubière cedex, France

**Bold, S.**
Center for Applied Geoscience, University of Tübingen, Sigwartstr. 10, D - 72076 Tübingen, Germany

**Bonneville, S.**
Department of Earth Sciences, Geochemistry, P.O. Box 80021, 3508 TA Utrecht, The Netherlands

**Bouchard, D.**
Center for Hydrogeology, University of Neuchâtel, Tue Emile-Argand 11, CH - 2007 Neuchâtel, Switzerland

**Breedveld, G.**
Norwegian Geotechnical Institute, Sognsveien 72, 0855 Oslo, Norway

**Broholm, M.**
Technical University of Denmark, Environment & Resources DTU, Building 115, DK - 2800 Kgs. Lyngby, Denmark

**BROUYÈRE, S.**
Hydrogeology and Environmental Geology, Dept of Georesources, Geotechnologies and Building Materials (GeomaC), University of Liège, Belgium, Building B52/3, 4000 Sart Tilman, Belgium

**CANDELA, L.**
Technical University of Catatonia – UPC, Dept. Geotechnical Engineering and Geosciences, Jordi Girona 1 - 3, Modulo D-2, Campus Nord 08028, Barcelona, Spain

**VAN CAPPELLEN, P.**
Department of Earth Sciences, Geochemistry, P.O. Box 80021, 3508 TA Utrecht, The Netherlands

**CAPRI, E.**
Università Cattolica del Sacro Cuore, Istituto di Chimica Agraria ed Ambientale, sezione Chimica Vegetale, Via Emilia Parmense 84, 29100 Piacenza, Italy

**COMBOURIEU, B.**
Dept. of Chemistry, Laboratoire Synthèse et Etude de Systèmes à Intérêt Biologique, UMR 6504 CNRS, University Blaise Pascal, 63117 Aubière cedex, France

**COMEGNA, A.**
Division of Water Resources Management, University of Naples Federico II, I - 80055 Portici (Naples), Italy

**DAHMKE, A.**
Institute of Geosciences, Dept. of Applied Geology, Christian Albrechts University Kiel, Ludewig-Meyn-Str. 10, D - 24118 Kiel, Germany

**DALE, A.W.**
Department of Earth Sciences, Geochemistry, P.O. Box 80021, 3508 TA Utrecht, The Netherlands

**DELORT, A.M.**
Dept. of Chemistry, Laboratoire Synthèse et Etude de Systèmes à Intérêt Biologique, UMR 6504 CNRS, University Blaise Pascal, 63117 Aubière cedex, France

**FANK, J.**
Joanneum Research, Institute for Water Resources Management - Hydrogeology and Geophysics, Elisabethstraße 16/II, A - 8010 Graz, Austria

**DI FAZIO, A.**
Dept. of Hydraulics, Transportation & Roads, University "La Sapienza", Via Eudossiana, 18, I - 00184 Rome, Italy

**FRAGOULIS, G.**
Università Cattolica del Sacro Cuore, Istituto di Chimica Agraria ed Ambientale, sezione Chimica Vegetale, Via Emilia Parmense 84, 29100 Piacenza, Italy

**GRANDEL, S.**
Institute of Geosciences, Dept. of Applied Geology, Christian Albrechts University Kiel, Ludewig-Meyn-Str. 10, D - 24118 Kiel, Germany

**GRATHWOHL, P.**
Center for Applied Geoscience, University of Tübingen, Sigwartstr. 10, D - 72076 Tübingen, Germany

**GRESKOWIAK, J.**
Leibniz-Institut of Freshwater Ecology and Inland Fisheries, Müggelseedamm 310, D - 12587 Berlin, Germany

**HAROUNE, N.**
Dept. of Chemistry, Laboratoire Synthèse et Etude de Systèmes à Intérêt Biologique, UMR 6504 CNRS, University Blaise Pascal, 63117 Aubière cedex, France

**HÖHENER, P.**
Swiss Federal Institute of Technology Lausanne (EPFL), CH - 1015 Lausanne, Switzerland

**HUNKELER, D.**
Center for Hydrogeology, University of Neuchâtel, Rue Emile-Argand 11, CH - 2007 Neuchâtel, Switzerland

**HYACINTHE, C.**
Department of Earth Sciences, Geochemistry, P.O. Box 80021, 3508 TA Utrecht, The Netherlands

**KEMNA, A.**
Institute Agrosphere, Dept. of Chemistry and Dynamics of the Geosphere, Forschungszentrum Jülich GmbH, D - 52425 Jülich, Germany

**KJELDSEN, P.**
Technical University of Denmark, Environment & Resources DTU, Building 115, DK - 2800 Kgs. Lyngby, Denmark

**KLONOWSKI, M.R.**
Polish Geological Insititute Lower Silesian Branch, Jaworowa 19, Pl-53-122 Wroclaw, Poland

**KNUDSEN, J.B.S.**
Norconsult, Vestfjordgaten 4, 1338 Sandvika, Norway

**KUZNETSOV, M.**
Ben-Gurion University of the Negev, J. Blaustein Institutes for Desert Research, Zuckerberg Institute for Water Research, Environmental Hydrology & Microbiology, Sde Boker Campus, 84990, Israel

**LAVERMAN, A.M.**
Department of Earth Sciences, Geochemistry, P.O. Box 80021, 3508 TA Utrecht, The Netherlands

**LECCESE, M.**
Dept. of Hydraulics, Transportation & Roads, University "La Sapienza", Via Eudossiana, 18, I - 00184 Rome, Italy

**LIEDL, R.**
Center for Applied Geoscience, University of Tübingen, Sigwartstr. 10, D - 72076 Tübingen, Germany

**van LITH, Y.**
Department of Earth Sciences, Geochemistry, P.O. Box 80021, 3508 TA Utrecht, The Netherlands

**MAIER, U.**
Center for Applied Geoscience, University of Tübingen, Sigwartstr. 10, D - 72076 Tübingen, Germany

**MAJONE, M.**
Dept. of Chemistry, University "La Sapienza", P.le Aldo Moro 5, I - 00185 Rome, Italy

**MANCONI, A.**
Dept. of Land Engineering, Università degli Studi di Cagliari, Italy

**MAQUEDA, C.**
Institute of Natural Resources and Agrobiology, IRNAS (CSIC), Apdo 1052, 41080 Seville, Spain

**MASSMANN, G.**
Free University of Berlin, Institute of Geosciences, Malteserstr. 74-100, 12249 Berlin, Germany

**MORILLO, E.**
Institute of Natural Resources and Agrobiology, IRNAS (CSIC), Apdo 1052, 41080 Seville, Spain

**NIR, S.**
Faculty of Agriculture, HUJI, Rehovot 76100, Israel

**NÜTZMANN, G.**
Leibniz-Institut of Freshwater Ecology and Inland Fisheries, Müggelseedamm 310, D - 12587 Berlin, Germany

**PALLADINO, M.**
Department of Agricultural Engineering, University of Naples Federico II, 80055 Portici (Naples), Italy

**PALLUD, C.**
Department of Earth Sciences, Geochemistry, P.O. Box 80021, 3508 TA Utrecht, The Netherlands

**PEKDEGER, A.**
Free University of Berlin, Institute of Geosciences, Malteserstr. 74-100, 12249 Berlin, Germany

**PETRANGELI PAPINI, M.**
Dept. of Chemistry, University "La Sapienza", P.le Aldo Moro 5, I - 00185 Rome, Italy

**PROMMER, H.**
Utrecht University, The Netherlands; c/o CSIRO Land and Water, Private Bag No. 5, Wembley WA 6913, Australia

**PUGLISI, E.**
Università Cattolica del Sacro Cuore, Istituto di Chimica Agraria ed Ambientale, sezione Chimica Vegetale, Via Emilia Parmense 84, 29100 Piacenza, Italy

**REGNIER, P.**
Department of Earth Sciences, Geochemistry, P.O. Box 80021, 3508 TA Utrecht, The Netherlands

**ROCK, G.**
JOANNEUM RESEARCH, Institute for Water Resources Management - Hydrogeology and Geophysics, Elisabethstraße 16/II, A - 8010 Graz, Austria

**ROMANO, N.**
Department of Agricultural Engineering, University of Naples Federico II, 80055 Portici (Naples), Italy

**RONEN, D.**
Water Quality Division, Israel Water Commission, 14 Hamasger St., POB 20365, Tel-Aviv 61203, Israel

**RODRÍGUEZ, R.**
Department of Geotechnical Engineering & Geosciences. School of Civil Engineering - UPC -. Gran Capitán s/n, Edif. D-2. Barcelona 08034, Spain

**ROSSETTI, S.**
Water Research Institute (IRSA-CNR), Via Reno 1, I - 00198, Rome, Italy

**RYTWO, G.**
Tel Hai Academic College, Upper Galilee 12210, Israel

**SANCELME, M.**
Dept. of Chemistry, Laboratoire Synthèse et Etude de Systèmes à Intérêt Biologique, UMR 6504 CNRS, University Blaise Pascal, 63117 Aubière cedex, France

**SANTINI, A.**
Department of Agricultural Engineering, University of Naples Federico II, 80055 Portici (Naples), Italy

**SCHÄFER, D.**
Dept. of Applied Geology, Christian-Albrechts-Universität zu Kiel, Olshausenstraße 40 - 60, D - 24098 Kiel, Germany

**SEVERINO, G.**
Division of Water Resources Management, University of Naples Federico II, I - 80055 Portici (Naples), Italy

**SERBAN, C.**
Faculty of Agriculture, HUJI, Rehovot 76100, Israel

**SOMMELLA, A.**
Division of Water Resources Management, University of Naples Federico II, I - 80055 Portici (Naples), Italy

**SOREK, S.**
Ben-Gurion University of the Negev, J. Blaustein Institutes for Desert Research, Zuckerberg Institute for Water Research, Environmental Hydrology & Microbiology, Midreset Ben-Gurion, Sde Boker Campus 84990, Israel

**STRACQUALURSI, N.**
Dept. of Civil Engineering - University "Tor Vergata", Rome Italy, V. di Tor Vergata 110, I - 00133 Rome, Italy

**TANDOI, V.**
Water Research Institute (IRSA-CNR), Via Reno 1, I - 00198, Rome, Italy

**THULLNER, M.**
Department of Earth Sciences, Geochemistry, P.O. Box 80021, 3508 TA Utrecht, The Netherlands

**TILLMANN, A.**
Institute Agrosphere, Dept. of Chemistry and Dynamics of the Geosphere, Forschungszentrum Jülich GmbH, D - 52425 Jülich, Germany

**TREVISAN, M.**
Università Cattolica del Sacro Cuore, Istituto di Chimica Agraria ed Ambientale, sezione Chimica Vegetale, Via Emilia Parmense 84, 29100 Piacenza, Italy

**UNDABEYTIA, T.**
Institute of Natural Resources and Agrobiology, IRNAS (CSIC), Apdo 1052, Avda. Reina Mercedes 10, 41080 Seville, Spain

**VANDERBORGHT, J.**
Institute Agrosphere, Dept. of Chemistry and Dynamics of the Geosphere, Forschungszentrum Jülich GmbH, D - 52425 Jülich, Germany

**VEREECKEN, H.**
Institute Agrosphere, Dept. of Chemistry and Dynamics of the Geosphere, Forschungszentrum Jülich GmbH, D - 52425 Jülich, Germany

**VERWEERD, A.**
Institute Agrosphere, Dept. of Chemistry and Dynamics of the Geosphere, Forschungszentrum Jülich GmbH, D - 52425 Jülich, Germany

**VIOTTI, P.**
Dept. of Hydraulics, Transportation & Roads, University "La Sapienza", Via Eudossiana, 18, I - 00184 Rome, Italy

**YAKIREVICH, A.**
Ben-Gurion University of the Negev, J. Blaustein Institutes for Desert Research, Zuckerberg Institute for Water Research, Environmental Hydrology & Microbiology, Sde Boker Campus, 84990, Israel

**ZHENG, Z.**
Ernest Orlando Lawrence Berkeley National Laboratory
Earth Science Division, MS: 70-108B,
One Cyclotron Road, Berkeley, CA 9472, USA

# 1 Processes

# The unsaturated zone – a neglected component of nature

D. Ronen[1, 2], S. Sorek[2, 3]

[1] Water Quality Division, Israel Water Commission, POB 20365, Tel-Aviv 61203, Israel
[2] Ben-Gurion University of the Negev, J. Blaustein Institutes for Desert Research, Zuckerberg Institute for Water Research, Environmental Hydrology & Microbiology, Sde Boker Campus, 84990, Israel
[3] Ben-Gurion University of the Negev, Mechanical Engineering, Pearlstone Center for Aeronautical Engineering Studies, Beer Sheva 84105, Israel

Corresponding author: 14 Hamasger St., POB 20365 – Tel Aviv 61203 – Israel – Tel: +972.3.6369648 – Fax: +972.3.9604017 – E-mail: danronen@bgu.ac.il

## Abstract

The unsaturated zone and the saturated/unsaturated interface region are important links between groundwater and the land surface. They provide storage capacity for both water and contaminants; a reactor medium for physical, chemical and biological processes; a delay time between the release of a contaminant into the unsaturated zone and its influx into the saturated zone and, a domain for the lateral and vertical transport of gases. We highlight some of the properties associated with these domains providing results of studies conducted in the phreatic Coastal Plain aquifer of Israel for the last 30 years.

## Keywords

Unsaturated zone; saturated/unsaturated interface region; nitrates, chlorides.

## Introduction

Understanding contaminant behavior in the unsaturated zone (UZ) and at the saturated/unsaturated interface region (SUIR) is a prerequisite for efficient management of groundwater resources. At the present level of knowledge the quantitative relationship between the amount of a contaminant released at the soil surface, e.g. by agricultural, or industrial and urban activity, and its resulting concentration in groundwater is uncertain. This is primarily due to lack of both knowledge and scarcity of data concerning the physical, chemical, biological and transport processes undergone by contaminants in the UZ and at the SUIR

The UZ and the SUIR are important links between groundwater and land surface. They provide: (1) storage capacity for both water and contaminants, (2) a delay time between the release of contaminants into the UZ and their influx into the saturated zone, (3) a reactor medium for physical, chemical and biological processes and, (4) a 3-D domain for the transport of gases and liquids.

In what follows we highlight some of the properties of the UZ and the SUIR on the basis of observations, conducted in the phreatic Coastal Plain aquifer of Israel for the last 30 years, as well as on modeling results. Namely, our idea is to draw attention to processes, fluxes and magnitudes of reservoirs, with the understanding that, even for the studied aquifer, some values presented are only preliminary estimates that disclose the complexity of the system.

## The Coastal Plain aquifer of Israel

The Coastal Plain aquifer stretches for 120 km south of Mount Carmel; its width varies between 7 and 20 km and the surface area is 2,000 km$^2$. It has a wedge-like cross section with a maximum thickness of 180 m near the sea that tapers towards the east until it disappears near the foothills of the Samarian and Judean Mountains. The aquifer is composed of clastic sediments (sand, sandstone, calcareous sandstone, siltstone and red loamy soils; Issar 1968). The thickness of the unsaturated zone ranges from 4 to

98 m, averaging 34 m (Mercado 1975). The aquifer is naturally replenished by rain during the winter months, October to March (average rain of 500 mm $yr^{-1}$). Intensive agricultural development of the replenishment area of the aquifer took place at the beginning of the $20^{th}$ century. Presently, major cities and industrial areas and more than 50% of Israel's total population reside on top of the Coastal Plain aquifer. The average concentration of chlorides ($Cl^-$) and nitrates ($NO_3^-$) in groundwater is increasing continuously as a result of agriculture, urban and industrial activities at an average rate, since the early 1970s, of 3 mg $Cl^-$/L yr (from 110 mg $Cl^-$/L in 1945 to 205 mg $Cl^-$/L in 2002) and 0.6 mg $NO_3^-$/L yr (from 45 mg $NO_3^-$/L in 1962 to 62 mg $NO_3^-$/L in 2002; Hydrological Service, 2003).

In 1980 it was estimated that the chloride influx to land surface (Fig.1) from rain, fallout, return flow and water import was of 106 Kt $Cl^-$/yr (kiloton chloride per year; Orenstein and Mercado 1980). In the early 1970's, it was calculated that 24 Kt N/yr is the net nitrogen input (Fig. 2) to the replenishment area of the aquifer (net denotes nitrogen amounts not utilized by vegetation; Ronen et al. 1983) from continuous contamination sources such as fertilizers and manure (58% of the net nitrogen input), sewage (16%), animal excreta (10%), irrigation water (7%), rain water (6%) and solid waste (3%). However, it was also postulated that most of the nitrates found in the aquifer reached the water table in the period 1930 to 1960, as a result of the oxidation of soil organic matter during reclamation of swamps and cultivation of virgin soils. It was estimated that this two exhaustive sources contributed about 1,100 Kt N during the period 1900 to 1930 (Ronen et al. 1983, 1984).

In some urban and industrial regions of the Coastal Plain, mainly in the Metropolitan Tel Aviv area, the aquifer is severely contaminated by Volatile Organic Compounds (Graber et al. 2002).

## The UZ – a large storage reservoir

The amount of water stored in the unsaturated zone (Fig. 1) was estimated to be 11 x $10^9$ $m^3$ (Mercado 1975, Orenstein and Mercado 1980). This is a considerable amount of water as compared to 21 x $10^9$ $m^3$ water contained in the upper active region of the aquifer (about 66% of the saturated thickness of the aquifer) where most pumping wells are situated. Obviously, the interstitial water in the UZ has a chemical signature. For example, under areas irrigated with sewage effluents the maximum concentrations of chloride, nitrate and Dissolved Organic Carbon (DOC) detected were 500 mg/L, 600 mg/L (Gvirtzman et al. 1986) and 100 mg/L (Amiel et al. 1990), respectively.

What could be the magnitude of the chloride pool and the nitrogen and organic carbon mass stored in the UZ of the aquifer?

Using a simulation model (COMPFLOW, Unger et al. 1996) and based on the management pattern of the aquifer during the last hundred years, Mercado (2001) estimated (Fig. 1) that the amount of $Cl^-$ stored in the UZ increased from 300 Kt $Cl^-$ in the year 1900 to 2,070 Kt $Cl^-$ in 1992. He also suggested that in the year 2015 this amount will increase to 3,105 Kt $Cl^-$. For the year 1980 Orenstein and Mercado (1980) estimated that the chloride flux into groundwater from the UZ was of 58.6 Kt $Cl^-$. At that time, the estimated amount of $Cl^-$ in the unsaturated zone was of 1,590 Kt $Cl^-$and in the saturated zone 3,150 Kt $Cl^-$. The significant amount of chloride in the unsaturated zone was referred to as "the chloride time-bomb".

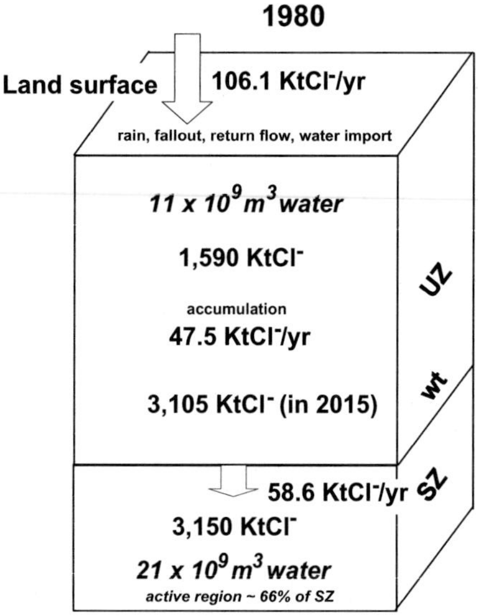


**Fig. 1.** Chloride pool and water pool in the unsaturated zone (UZ) of the Coastal Plain aquifer of Israel in the 1980's. The water pool in the saturated zone is calculated for the upper active region of the aquifer (66% of the saturated thickness of the aquifer) where most pumping wells are located. The upper arrow denotes chloride influx to land surface from rain, fallout, return flow and water import. The lower arrow denotes chloride flux from the UZ into the saturated zone (SZ). Also shown is the chloride pool in the SZ; wt - water table (after Mercado et al. 1975, Orenstein and Mercado 1988, Mercado 2001). According to the simulation model the amount of $Cl^-$ stored in the UZ increased from 300 Kt $Cl^-$ in 1900 to 2,070 Kt $Cl^-$ in 1992. In 2015 this amount will increase to 3,105 Kt $Cl^-$.

Based on a series of 39 profiles from research wells representing most of the lithologic units of the aquifer and located in areas characterizing most current land uses (virgin soils, non-irrigated and irrigated crops, citrus groves, effluent irrigated areas, barn yards, sewage and solid waste disposal sites) Raveh (1972) estimated that the total amount of nitrogen stored in the UZ in the year 1970 was of 9,560 Kt N. Out of this amount 200 Kt N are present as $NO_3^-$ (N-$NO_3^-$; Fig 1) and 160 Kt N as $NH_4^+$ (N-$NH_4^+$), the rest is organic nitrogen (N-NOrg). For the year 1970 Ronen et al. (1984) estimated that the input of nitrogen to the saturated zone was 1.3 Kt N and the amount of $NO_3^-$ stored in groundwater 100 Kt N. Considering the calculated N/C ratios (Ronen et al. 1984) it was also estimated that the total amount of organic carbon (C-Org) in the UZ is of 105,000 Kt C-Org (Fig. 2).

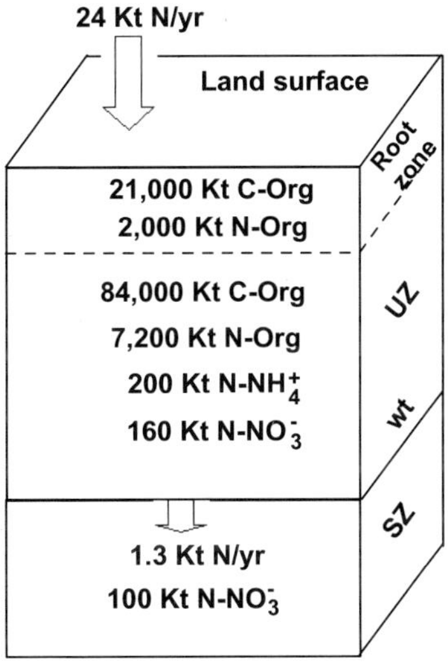


**Fig. 2.** Nitrogen and carbon pools in the unsaturated zone (UZ) of the Coastal Plain aquifer of Israel in the early 1970's. The upper arrow denotes net nitrogen influx to land surface from natural and anthropogenic sources. The lower arrow shows nitrogen flux from the UZ into the saturated zone (SZ). Also shown is the nitrogen pool in the SZ; wt – water table (after Ronen et al. 1984)

The nitrogen reservoir in the UZ is orders of magnitude higher than the nitrogen reservoir found in the saturated zone (most of it as $NO_3^-$) and the nitrogen fluxes to the soil surface and the saturated zone. Clearly, the organic matter in the UZ is also a potential source of both $NH_4^+$ and $NO_3^-$. On the other hand, the anaerobic decomposition of organic matter may lead to the reduction of nitrate.

## The UZ and SUIR – a time-buffer zone

Calculations based on the aquifer replenishment and the water content of the UZ suggest that the average transit time of water to the water table is 27 years ranging from 1 to 50 years in most of the surface area of the Coastal Plain aquifer of Israel. For some restricted regions higher values to the extent of 800 years were calculated (Mercado 1975).

A detailed study where water in the UZ was traced according to its tritium content, utilizing the difference between the environmental tritium content of rain and irrigation water (Gvirtzman et al. 1986), demonstrated that in a clay loam layer the average vertical velocity is about 0.7 m/yr and in sandy sediments the mean rate of vertical water movement is 2.3 m/yr. The same study provided evidence of anion exclusion, where in the clay loam layer the vertical velocity of $Cl^-$ and $SO_4^=$ was 1.35 m/yr.

An additional delay factor between land surface and the saturated zone is imposed by the SUIR. Ronen et al. (2000) calculated that for a SUIR having a thickness of 3 m, the residence time of recharge water may be greater than 5 years.

Based on these estimates, and considering retardation factors for chemical moieties, it can be suggested that, for the Coastal Plain aquifer of Israel, decades may elapse from the time that a contaminant is released at the replenishment area of the aquifer until the moment it reaches the saturated zone.

## The SUIR – a singular biochemical reactor

The SUIR extends from an imaginary plane of 100% water saturation and positive water pressure ($p > 0$) to the surface of the capillary fringe (CF; Fig. 3a), the imaginary plane where $p < 0$ and water content ($\theta$) is equal to the background residual water content of the UZ (Ronen et al. 2000). The SUIR is generally characterized by high pore water content as compared with residual pore water in the unsaturated zone, and by a relatively large gas-filled porosity as compared to that in the saturated zone (Fig. 3a; Ronen et al. 1997 and 2000). Moreover, tortuosity and the area available for liquid and vapor mass fluxes at the SUIR are both highly variable (as compared to the UZ proper) since moisture content is also spatially and temporally variable (Ronen et al. 2000). For example, in a study where 13 boreholes were located in relatively homogeneous sandy sediments within a radius of 5 m, the height of the CF (about 1.4 m) was found to vary by up to 50% (Ronen et al. 2000).

Field evidence indicates that high concentrations of DOC (DOC 20 to > 100 mg/L; Fig. 3b) are found throughout the 30 m thick UZ under land irrigated with sewage effluents since the early 1960's (Amiel et al. 1990). The study area, where bulk groundwater contains high Dissolved Oxygen concentrations (DO ~ 7.5 mg/L), is under a high organic carbon load. About 14 gr C-Org/$m^2$ were annually applied each summer, in contrast to the negligible amounts added to fields irrigated with fresh waters. DOC mobility is inferred from the lack of correlation between the content of DOC and Total Organic Carbon (TOC) in the sediments of the UZ (Amiel et al. 1990). The DOC is not completely biodegraded as it percolates from the soil surface to groundwater. A DOC mass balance indicates that the unsaturated zone still contains about 50% of the total DOC input. DOC persistence for more than 15 years under unsaturated conditions suggests that moisture content may be a significant controlling factor for biodegradation. Indeed, at the SUIR, where water content increases significantly, part of the DOC is biodegraded (Fig. 3b) and anoxic conditions develope (DO < 1 mg/L, Fig. 3c Ronen et al. 1987). In some cases the system becomes anaerobic and denitrification follows (Fig. 3d; Ronen et al. 1987).

Still to be answered is the question as to whether anoxification at the SUIR could be expected in other areas where fresh water is utilized for irrigation in the Coastal Plain aquifer of Israel. Few SUIR research wells are available in the aquifer, but the answer to this question may be inferred from the high oxygen content of groundwater as found in the deep pumping wells of the aquifer (Ronen et al. 1987). These pumping wells penetrate about two thirds of the saturated thickness of the aquifer. The residence time of water in this portion of the aquifer is about 25 years. As the average seepage time of water through the UZ is about 30 years, the present average oxygen content of groundwater in the pumping wells reflects the replenishment conditions in the 1940's. The intensive agricultural development of the Coastal Plain region started at the beguinning of the $20^{th}$ century. If we consider retardation in the transport of DOC, it may be postulated that the organic matter, mobilized as the result of the onset of the agricultural activity (Ronen et al. 1984 ), has only lately reached the saturated zone. As explained before: (a) it was suggested that the massive contamination of groundwater by nitrates has resulted from the oxidation of soil organic matter due to the reclamation of swamps and the cultivation of virgin soils at the beguinning of the $20^{th}$ cetury (Kanfi et al. 1983), and (b) the rate of percolation of anions through the unsaturated zone may be even faster than that of water (Gvirtzman et al. 1986). Therefore, nitrates may be expected to have reached groundwater earlier than the water soluble organic carbon released from the same source. Hence, the assumption that an anaerobic layer has been developed in large regions of the SUIR of the

Coastal Plain aquifer of Israel may not be dismissed on the basis of the present available data. The 95% decrease between nitrogen fluxes to land surface and to the saturated zone (Fig. 2) may be the result of denitrification at the water table region (Fig. 3d).

The biodegradation of DOC, under land irrigated with sewage effluents, leads to the production of gases in porous media. In air samples extracted from the SUIR, the concentration of $CO_2$ was up to 2% (Fig 3e; Affek et al. 1998) and the annual $CO_2$ flux from the SUIR towards the atmosphere (6.3 g C/m$^2$ yr) is balanced by a similar influx of DOC from the sewage effluents. At the SUIR, the $\delta^{13}C$ of $CO_2$ indicates that it is produced from the biodegradation of sedimentary organic carbon, probably as a result of competitive sorption (i.e., exchange between DOC and sedimentary organic carbon; Affek et al. 1998).

Nitrous oxide ($N_2O$) was also found to be produced at the SUIR and the concentrations detected in the liquid phase of the SUIR were up to 400 µg/L (Fig. 3f; Ronen et al. 1988) about three orders of magnitude higher than the concentration expected from equilibrium with the earth's atmosphere. In some cases the production of $N_2O$ by nitrification may be inferred from the very high concentrations of $N_2O$ found, and the concomitant decrease in pH and bicarbonate content which may be the result of nitrogen mineralization and cell synthesis of nitrifying organisms, respectively.

The biologically-produced gases may persist as gas bubbles in groundwater. Bubbles clog pores and therefore reduce the hydraulic conductivity without significantly reducing the volumetric water content (Ronen et al. 1989). This phenomenon is the result of the very high pressures required to force a bubble through a pore space and to overcome the resistance to flow offered by detached gas bubbles and liquid drops in capillary conduits. A second mechanism by which bubbles may be produced at the SUIR is during recharge. Entrapped air is one of the reasons for the hysteretic relationship between water content and pressure head during drainage and imbibition.

Calculations (Ronen et al. 1989) show that the critical depth below the water table ($p = 0$) at which bubbles are most likely to be found in the Coastal Plain aquifer of Israel, is of about 1 m. In the same aquifer, unsaturated conditions were detected below the water table till a depth of about 1.5 m (Fig. 3a) below the water table (Ronen et al. 2000). This estimate coincides with the depth of 0.60 m of an almost stagnant water layer found in an area where groundwater is driven by natural gradient flow conditions. The specific discharge was found to increase with depth from 0.5 m/yr (water velocity ~ 1 m/yr), near the water table, to 4.5 m/yr (water velocity ~ 10 m/yr) at a depth of 2.4 m below the water table (Fig. 3g; Ronen et al. 1986) and up to 20 m/yr (water velocity ~ 44 m/yr) at a depth of 7 m.

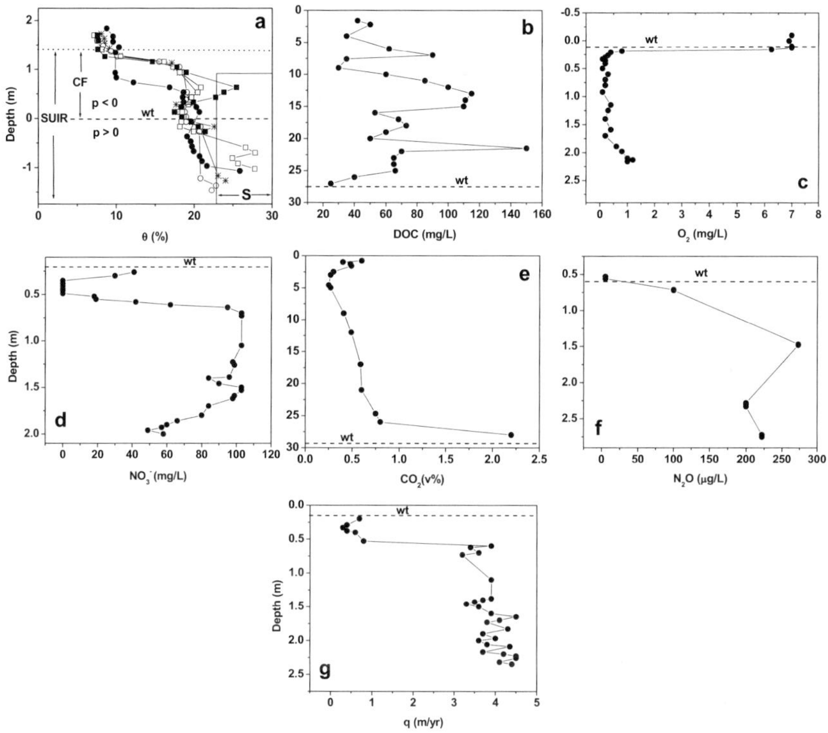


**Fig. 3.** Profiles obtained in the UZ and the SUIR of the Coastal Plain aquifer of Israel: (a) gravimetric water content (θ) in the SUIR. The inner frame denotes saturated conditions (S) at the SUIR, below and above the water table (wt); CF – capillary fringe (after Ronen et al. 2000; J of Contaminant Hydrology); (b) DOC in the UZ and CF (after Amiel et al. 1990; Research Journal WPCF); (c) anoxification in the SUIR (after Ronen et al. 1987, WRR); (d) profile showing denitrafication just below the water table (after Ronen et al. 1987, Journal of Hydrology); (e) $CO_2$ production at the CF (after Affek et al. 1998, WRR); (f) production of $N_2O$ (Ronen et al. 1988, Nature); (g) stagnant conditions that develop due to bubble production and air entrapment (after Ronen et al. 1986, WRR).

## UZ - a 3-D domain for transport of gases

Exceptionally high concentrations of volatile organic compounds (VOCs), e.g. Trichloroethylene (TCE) up to 124,000 μg/L-air in the unsaturated zone and up to 260,000 μg/L-water in the saturated zone, were measured at the SUIR in the metropolitan Tel Aviv area (at the Coastal Plain aquifer of Israel) where the land-atmosphere interface is characterized by large impermeable areas associated with urban infrastructure (Graber et al. 2002). Numerical simulations suggest that VOCs are being transported also in the gas phase of the unsaturated zone. Considering a single point source of TCE with a discharge rate of 0.2 mm/day, for 50 years, it was shown that TCE vapors may spread over an area of about 9 $km^2$ (Fig. 4). Vapor phase transport in the UZ is not related to the direction of water flow in the UZ or saturated zones. Thus, TCE vapors were found to become a secondary contamination source upstream of the single point source. Moreover, it was found that TCE vapors experience an upward flux at vertical impervious boundaries such that underground structures may be surrounded by VOC vapors. Indeed, VOCs were detected inside many buildings; the highest concentrations of TCE were 628 $\mu g/m^3$ air (Graber et al. 2002).

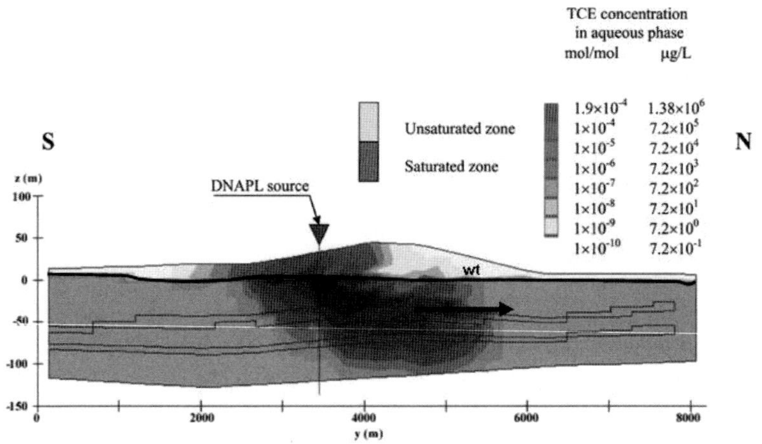


**Fig. 4.** Cross section of the Coastal Plain aquifer of Israel at the Tel Aviv area showing for the year 2000 the simulated distribution of TCE from one DNAPL source after a 50 years discharge rate of 0.2 mm/day. The TCE in the gaseous phase spreads in the UZ over an area of 9 $km^2$ also opposite to the direction of groundwater flow (arrow); wt – water table (after Graber et al. 2002).

## Conclusions

The mass of a chemical component stored in the UZ of an aquifer may be considerable as compared to inputs from natural and anthropogenic sources. In the Coastal Plain of Israel, in the year 1970 for example, the total amount of nitrogen in the UZ was 398 times higher than the yearly net nitrogen input to land surface from all nitrogen contamination sources, and 7,353 times higher than the yearly N-$NO_3^-$ influx to groundwater. In this same aquifer if all chloride contamination sources in 1980 were removed, the $Cl^-$ pool in the UZ would still continue to contaminate groundwater, with a similar chloride flux, during 27 years.

The SUIR is also an active interface that governs influx amounts reaching bulk groundwater and chemical speciation decades after contaminant penetration into the UZ. In some cases components of UZ sediments (e.g., sedimentary organic carbon) interact with interstitial water components playing an active roll in defining the chemical and isotopic composition of recharge.

The UZ operates also as a conduit for gases in urban and rural environments. The existence of industrial contamination sources in urban areas and the impervious boundary conditions that develop there create a unique problem that is not only related to groundwater quality but to the deterioration of the human habitat located within and above the UZ.

Obviously, the UZ chemical pools must be considered when developing managerial strategies for the prevention of aquifer contamination or for groundwater restoration. But do we know how to do it? For example, can we suggest reduction of fertilizer and manure inputs in the Coatal Plain aquifer of Israel, claimed to be responsible for an yearly input of 58% of all nitrogen contamination sources (14 Kt N), while the nitrogen pool in the UZ is almost 10,000Kt N? At this stage there is no clear answer to this dilemma.

In spite of all mentioned attributes the UZ of the Coastal Plain aquifer of Israel, and probably the UZ of many aquifers of the world, continue to be "no-man's land". This is largely the result of: (1) the inherent high cost of monitoring techniques in a highly heterogeneous system, (2) the difficulty in convincing water managers about the importance of a zone from which it is impossible to pump and supply water, and (3) the perception of decision makers that, in many cases, this zone will define the future of water resources and urban habitats only decades after their administrative cadence. Available evidence suggests that this may be a naive approach leading to irreversible degradation of aquifer and groundwater quality.

## References

Affek HP, Ronen D, Yakir D (1988) About production of $CO_2$ in the capillary fringe of a deep phreatic aquifer. Water Resources Research, Vol. 34, pp. 989-996

Amiel AJ, Magaritz M, Ronen D, Lindstrand O (1990) Dissolved organic carbon in the unsaturated zone under land irrigated by sewage-effluents. Research Journal Water Pollution Control Federation, Vol. 62, pp. 861-866

Graber ER, Ronen D. Elhanany S. Laor L, Sorek S, Yakirevich A (2002) Assessment of Aquifer Contamination in the Nahalat Itzhak Area – Tel Aviv. Final Report to the Israel Water Commission, (Vol. I, II and III)

Gvirtzman H, Ronen D, Magaritz M (1986) Anion exclusion during transport through the unsaturated zone" Journal of Hydrology, Vol. 87, pp. 267-283

Hydrological Service, Israel Water Commission "Development and exploitation of Israel water resources till autumn 2002", 2003, 300 p. (in Hebrew).

Issar A (1968) Geology of the central Coastal Plain of Israel. Israel Journal of Earth-Science, Vol. 17, pp. 16-29

Kanfi Y, Ronen D, Magaritz M (1983) Nitrate trends in the Coastal Plain Aquifer of Israel. Journal of Hydrology, Vol. 66, pp. 331-341

Mercado A (2001) Selected Groundwater Quality Parameters: Observed Trends and Possible Remedies. in Efficient Use of Limited Water Resources, Making Israel a Model State, Edit: Barry Rubin, Begin Sadat (BESA) Center For Strategic Studies, Bar Ilan University, pp. 29-42

Mercado A, Evron M, Kahanovitch I (1975) Salinity of groundwater in the Coastal Plain. TAHAL, Interim Report No. 1, 68 p. (in Hebrew)

Orenstein I, Mercado A (1988) Coastal Plain aquifer – quantity and quality aspects. TAHAL, Interim Report 1, 56 p. (in Hebrew)

Rave I (1972) Distribution of nitrogen compounds in the unsaturated zone of the Coastal Plain in Israel. M.Sc. Thesis, Technion – Israel Institute of Technology, 149 p. (in Hebrew)

Ronen D, Kanfi Y, Magaritz M (1983) Sources of nitrates in groundwater of the Coastal Plain of Israel - Evolution of ideas. Water Research, Vol. 17, 1499-1503

Ronen D, Kanfi Y, Magaritz M (1984) Nitrogen presence in groundwater as affected by the unsaturated zone" in Pollutants in Porous Media. Edited by B. Yaron, G. Dagan and J. Goldshmid, Springer, Verlag, pp. 223-236

Ronen D, Magaritz M, Paldor N, Bachmat Y (1986) The behavior of ground-water in the vicinity of the water table evidenced by specific discharge profiles. Water Resources Research, Vol. 22, pp. 1217-1224

Ronen D, Magaritz M, Almon E, Amiel H (1987) Anthropogenic anoxification ("eutrofication") of the water table region of a deep phreatic aquifer. Water Resources Research, Vol. 23, p. 1554-1560

Ronen D, Magaritz M, Gvirtzman M, Garner W (1987) Microscale chemical heterogeneity in groundwater. Journal of Hydrology, Vol. 92, pp. 173-178

Ronen D, Magaritz M, Almon E (1988) Contaminated aquifers are a forgotten component of the global $N_2O$ budget. Nature, Vol. 335, pp. 57-59

Ronen D, Berkowitz B, Magaritz M (1989) The development and influence of gas bubbles in phreatic aquifers under natural flow conditions. Transport in Porous Media, Vol. 4, pp. 295-306

Ronen D, Scher H, Blunt M (1997) On the structure and flow processes in the capillary fringe of phreatic aquifers. Transport in Porous Media, Vol. 28, pp. 159-180

Ronen D, Scher H, Blunt M (2000) Field observations of a capillary fringe before and after a rainy season. Journal of Contaminant Hydrology, Vol. 44, pp. 103-118

Unger AJA , Forsyth PA, Sudicky EA (1996) Variable spatial and temporal weighting schemes for use in multi-phase compositional problems. Adv. Water Resour., Vol. 19, pp. 1-27

# Use of stable isotope analysis to assess biodegradation of petroleum hydrocarbons in the unsaturated zone. Laboratory studies, field studies, and mathematical simulations

D. Bouchard[1], D. Hunkeler[1], P. Höhener[2], R. Aravena[3], M. Broholm[4], P. Kjeldsen[4]

[1] Center for Hydrogeology, University of Neuchâtel, Rue Emile-Argand 11, CH-2007 Neuchâtel
[2] Swiss Federal Institute of Technology Lausanne (EPFL), CH-1015 Lausanne
[3] Department of Earth Sciences, University of Waterloo, 200 University Avenue West, Waterloo, Ontario, Canada
[4] Technical University of Denmark, Environment & Resources DTU, Building 115, DK-2800 Kgs. Lyngby,

Corresponding author: Center for Hydrogeology, University of Neuchâtel, Rue Emile-Argand 11, CH-2007 Neuchâtel, Tel. +41 32 718 25 60, Fax + 41 32 718 26 03, Daniel.Hunkeler@unine.ch

## Abstract

Compound-specific isotope analysis is increasingly used to demonstrate contaminant degradation in groundwater. The method relies on the frequent occurrence of characteristic shifts in the isotope ratios of contaminants due to the faster degradation of molecules with light isotopes compared to those with heavy isotopes. The goal of the study was to evaluate if the method can also be used to assess biodegradation of petroleum hydrocarbons in the unsaturated zone. The study included laboratory experiments to determine isotopic enrichment for biodegradation of petroleum hydrocarbon compounds under unsaturated conditions and a field experiment. The field experiment consisted of burying an artificial fuel source in the unsaturated zone at a site in Denmark. Concentration and isotope ratios of individual compounds were monitored using a dense network of sampling points. Significant shifts of the isotope ratios of most of the compounds occurred. Initially, a depletion in $^{13}C$ with distance was observed likely due to the faster diffusion of molecules with $^{12}C$ only. Later, most of the compounds became in enriched in $^{13}C$ compared to the source due to biodegradation. To evaluate the relative contribution of diffusion and biodegradation in more detail, the concentration and isotope ratio evolution was simulated using an analytical model. The calculations confirmed that diffusion can lead to significant isotope fractionation at early times. For hydrogen isotopes, the effect of diffusion relative to biodegradation is expected to be smaller due to the large hydrogen isotope fractionation frequently observed during biodegradation of organic compounds. In conclusion, the study demonstrates that especially at early times after spills, the effect of diffusion has to be taken into account in the interpretation of isotope data from the unsaturated zone. Furthermore, it demonstrates that for many compounds, hydrogen isotopes are likely the more sensitive indicator of biodegradation.

## Keywords

Biodegradation; Petroleum hydrocarbons; stable isotopes; unsaturated zone

## Introduction

Natural attenuation is an attractive remediation strategy when dealing with petroleum hydrocarbon-contaminated sites because of its cost efficiency. Biodegradation is usually the main process leading to contaminant elimination and is usually considered to be the only process to influence $^{13}C/^{12}C$ and D/H ratio of organic contaminants in the saturated zone. Therefore, carbon and hydrogen isotope analysis has been used as a tool to demonstrate biodegradation (Griebler et al. 2004, Steinbach et al. 2004). Carbon and hydrogen isotope fractionation occurs during biodegradation as a consequence of the slightly faster cleavage of chemical bonds between light isotopes of an element. The difference in degradation rates leads to an enrichment of $^{13}C$ in the residual contaminant pool compared to the initial value. The degree of isotope fractionation generally varies for different compounds and/or degradation pathways and is usually expressed as isotope enrichment factor (Clark and Fritz 1997). Isotope enrichment factors for aerobic biodegradation have been determined for only few petroleum hydrocarbons compounds so far. Most of the field studies focused on the saturated zone (Meckenstock et al. 2004) compare to only few studies on the unsaturated zone (Kirtland et al. 2005, Stehmeier et al. 1999). The aim of this study was to evaluate whether compound-specific stable isotope analysis can be used to demonstrate biodegradation of petroleum hydrocarbons in the unsaturated zone. The study included the determination of isotopic enrichment factors for various hydrocarbons compounds in unsaturated microcosms as well as a field study at an experimental site. At the field site, a defined mixture of hydrocarbons was buried in a sandy unsaturated zone and the evolution of concentration and isotope ratios of various hydrocarbons was followed using a dense network of sampling points. The study was complemented with mathematical simulations performed to gain insight factors controlling isotope ratios in the unsaturated zone. To realize the simulations, the isotopic enrichment factors from the laboratory study and biodegradation rate from the field experiment were used along with physicochemical and geological parameters. Because the main transport of VOCs in the unsaturated zone is by diffusion through soil air and because $^{12}$C-compounds diffuse faster than $^{13}$C-containing compounds, isotope fractionation due to diffusion was taken into account in addition to biodegradation.

## Materials and methods

### Microcosm

A mixture containing *n*-hexane, *n*-octane, benzene and toluene was prepared to evaluate the carbon isotope enrichment factor during aerobic biodegradation. The molar ratio of compounds in the mixture was adjusted according to Raoult's law in order to have the same vapor pressure of each compound in the gas phase above the liquid. In the same way, a two-compound mixture made of methylcyclopentane and methylcyclohexane, a two-compound mixture made of *n*-hexane and *n*-pentane were also prepared and 3-methylpentane was tested as a pure compound. 2,2,4-trimethylpentane was used as non degradable reference compound. To simulate unsaturated aerobic conditions, 60 ml bottles were filled with alluvial sand containing a microbial community which had been shown previously to biodegrade petroleum hydrocarbons (Höhener et al. 2003). The sand was sieved < 4mm and fertilized with ammonium nitrate and potassium phosphate at a C:N:P ratio of 1:25:250. Total porosity of the soil was 0.41 with a water content of 5% (w/w) and an organic carbon content of 0.2% (Pasteris et al. 2002). No microorganisms were added to the indigenous population. Microcosms were closed with Teflon Mininert valves. 10 ml of vapour from the headspace of the hydrocarbon mixture was added to the microcosm. Degradation was monitored in air samples which were taken to determine concentrations and to quantify carbon isotope fractionation from the remaining compound. To account for possible effects of gas-aqueous phase partitioning and adsorption on $\delta^{13}C$ values, the same procedures as described previously was also carried out with sterile controls. The soil had been autoclaved three times within three days, each time for 20 minutes at a temperature of 120 $^0$C. Prior to filling the soil into the bottle, 3 ml of $NaN_3$ was added to soil to insure total inhibition of microbial activity.

Air samples were taken from the microcosm with a gastight syringe and were analysed using a TRACE™ gas chromatography coupled to an ThermoFinnigan™ Delta Plus XP isotope-ratio mass spectrometer via a ThermoFinnigan™ GC combustion III interface. Isotope ratios are reported using $\delta^{13}C$ notation relative to the VPDB standard (Clark and Fritz 1997) according to:

$$\delta C = \left( \frac{R_{sample}}{R_{reference}} - 1 \right) 1000 \tag{1}$$

Where $R_{sample}$ and $R_{reference}$ is the ratio of $^{13}C/^{12}C$ of the measured sample and the VPDB standard, respectively. Isotope fractionation factor was quantified using the Rayleigh type evolution model given by:

$$R_{sample} = R_0 \left( \frac{C_t}{C_0} \right)^{\alpha - 1} \tag{2}$$

Rearrangement of 2.2 and using the δ notation gives:

$$\ln \left( \frac{\delta^{13}C_{sample} + 1000}{\delta^{13}C_0 + 1000} \right) = (\alpha - 1) \ln \frac{C_{sample}}{C_0} \tag{3}$$

Where $\delta^{13}C_0$ is the initial isotope ratio, $\alpha$ is the isotope fractionation factor, $C_0$ and $C_{sample}$ are the initial concentration and remaining concentration at time *t*, respectively. The isotope fractionation factor $\alpha$ can be found using linear regression, where $\alpha-1$ is the slope of the curve. The isotope fractionation factor can be converted to isotope enrichment factor using:

$$\varepsilon = (\alpha - 1)1000 \quad (4)$$

## Field experiment

The field experiment was carried out at Værløse, Denmark as part of the GRACOS project. The experimental procedure for the field test consisted of burying a source of petroleum hydrocarbons made of known volatile organic compounds between 0.8 and 1.3 m depth in the unsaturated zone of a sandy aquifer (Christophersen et al. 2005). Soil gas concentrations were monitored in great detail using a dense network of sampling points. For this study, soil gas samples were then taken to analyze carbon isotope ratios of each petroleum compound at different distances from the source. Stable carbon isotope analyses were performed at the University of Waterloo using a Hewlett-Packard gas chromatograph connected via a combustion interface to a Micromass Isochrom isotope-ratio mass spectrometer (Micromass, Manchester, U.K.).

## Simulations

The mathematical simulations modeled two scenarios with either a constant or a decaying source located in the unsaturated zone. *N*-hexane was selected as a model compound. The *n*-hexane vapors diffuse through the homogeneous and isotropic porous media. The model included partitioning into the immobile soil water phase and biodegradation through a first-order reaction. The evolution of the $^{13}C/^{12}C$ ratio is modeled taking into account fractionation due to biodegradation and diffusion. Consequently, the results provide insights on the magnitude of fractionation by diffusion on the observed isotopic shifts in VOCs in the unsaturated zone. In order to reproduce the isotopic evolution, carbon and hydrogen enrichment factors determined previously in laboratory experiments along with first order biodegradation rate determined from inverse modeling of the field experiment (Gaganis et al. 2004) were used for the computation of the remaining concentration over time. Transport of gaseous compounds by diffusion and first-order degradation can be described by the following equation (Werner and Höhener 2003):

$$\partial/\partial t \, C_a = f_a \tau D_m \cdot \Delta C_a - k_{app} C_a \quad (5)$$

where $C_a$ is the n-hexane concentration in the soil air, $f_a$ is the fraction in the air, $\tau$ the tortuosity factor, $\Delta$ stands for the Laplace operator, $k_{app}$ is the apparent first-order degradation constant in the gas phase, and $D_m$ the molecular diffusion coefficient. For a continuous source with a radius $r_0$, a constant concentration $C_0$ at the source and an initial concentration of 0 everywhere else, the following solution of 2.5 is obtained:

$$C_a(r,t) = \frac{C_0 e^{-st} r_0}{2r}\left[ e^{-\sqrt{\frac{(k_{app}-s)}{D'}}(r-r_0)} erfc\left(\frac{(r-r_0)-2\sqrt{D'(k_{app}-s)}\,t}{2\sqrt{D't}}\right) + e^{\sqrt{\frac{(k_{app}-s)}{D'}}(r-r_0)} erfc\left(\frac{(r-r_0)+2\sqrt{D'(k_{app}-s)}\,t}{2\sqrt{D't}}\right)\right] \quad (6)$$

Where s is the source decay rate and $D' = f_a \tau D_m$.

### *Subspecies*

To be able to take into account differences in the biodegradation rate and diffusion velocity, n-hexane was subdivided into three different subspecies with a different isotope distribution (Table. 1):

- Subspecies I: Molecules with $^{12}C$ only
- Subspecies II: Molecules with one $^{13}C$ at any non-terminal position (position C2, C3, C4 or C5)
- Subspecies III: Molecules with one $^{13}C$ at the terminal positions (position C1 and C6)

Molecules with more than one $^{13}C$ were neglected since they are very rare, given a natural abundance of $^{13}C$ of only 1.11%. Slower diffusion is expected for all molecules with heavy isotope at any position (subspecies II and III, Table 1). However, slower biodegradation is only expected for subspecies III, since the initial attack occurs at a terminal position (Atlas 1981) and since shifts of the isotope ratio of the substrate reflect isotope fractionation associated with the initial step only. For hydrogen isotope fractionation, a similar grouping into subspecies is required (Table 1):

- Subspecies I: Molecules with H only
- Subspecies II: Molecules with one deuterium within the molecule (position C2, C3, C4 or C5)
- Subspecies III: Molecules with one deuterium at a terminal position (position C1 or C6)

As in the case for carbon, subspecies with more than one D are neglected.

**Table 1.** Decomposition of n-hexane in three subspecies (I, II and III), according to the isotope distribution and its influence on biodegradation rates (Kapp,L and Kapp,H) and diffusion coefficients (DL and DH).

| # | Carbon | Hydrogen | Degrad. rate | Diffusion Coeff. |
|---|---|---|---|---|
| I | 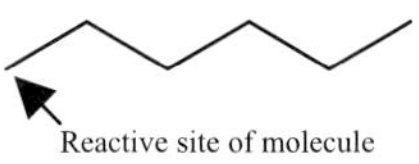  | 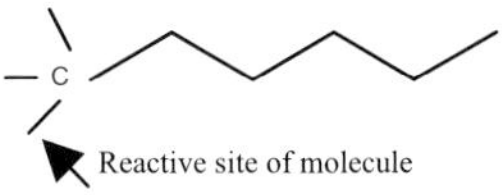  | Normal ($K_{app,L}$) | Normal ($D_L$) |
| II | 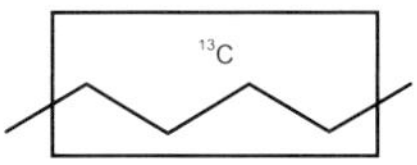  | 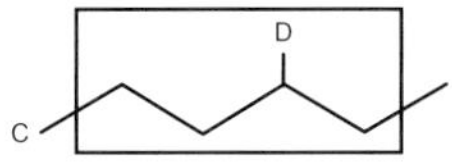  | Normal ($K_{app,L}$) | Slower ($D_H$) |
| III | 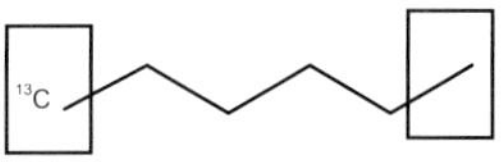  | 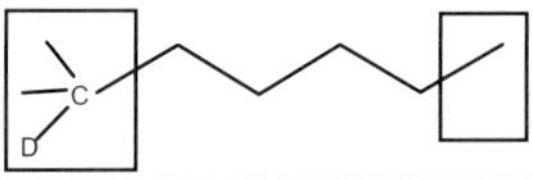  | Slower ($K_{app,H}$) | Slower ($D_H$) |

Once the 3 subspecies distinguished, an initial quantity is attributed to each of them. The total initial concentration of n-hexane was calculated based on the vapor pressure and mole fraction of n-hexane using Raoult's law. Based on the total concentration of carbon as hexane at the source, the concentration of $^{12}C$ and $^{13}C$ is calculated using equation 7 and 8:

$$C_0 = (12C_{12}) + (13C_{13}) \tag{7}$$

$$R_0 = \frac{C_{13}}{C_{12}} \tag{8}$$

where $C_0$ is the total initial concentration of carbon as n-hexane at the source, $C_{12}$ and $C_{13}$ is the concentration of $^{12}C$ and $^{13}C$, respectively, and $R_0$ is the initial isotopic ratio. By combining these two equations, $C_{12}$ and $C_{13}$ can be determined.

$$C_{12} = \frac{C_0}{(12 + R_0)13} \tag{9}$$

$$C_{13} = \frac{C_0}{\left(\frac{12}{R_0}\right) + 13} \tag{10}$$

Based on $C_{12}$ and $C_{13}$, the initial concentration of subspecies I ($C_{0,I}$), II ($C_{0,II}$) and III ($C_{0,III}$) are calculated assuming that $^{13}C$ is statistically distributed over all positions.

$$C_{0,I} = \frac{C_{12} - (C_{13}(n-1))}{n} \tag{11}$$

$$C_{0,II} = C_{13}\left(\frac{n-x}{n}\right) \tag{12}$$

$$C_{0,III} = C_{13}\left(\frac{x}{n}\right) \tag{13}$$

Where n is the number of carbon in the molecule, x is the number of carbon at terminal position. Note that subspecies II and III contain one $^{13}C$ per molecule. The concentration of subspecies II corresponds simply to the number of $^{13}C$ at intermediate positions, the concentration of subspecies III, to the number of $^{13}C$ at terminal positions. Subspecies I contains all $^{12}C$ except the $^{12}C$ of subspecies II and III that needs to be subtracted from the total quantity of $^{12}C$. The same types of calculations are made to obtain the initial moles of each subspecies for the H simulations.

**Biodegradation rates.** Two biodegradation rates were derived and attributed to the subspecies as follows: Biodegradation of species I and II proceeds at the overall rate ($k_{app,L}$), whereas for subspecies III, biodegradation is expected to be slower than for subspecies I and II. The biodegradation rate of subspecies III is related to that of the other subspecies by the fractionation factor specific for the terminal position $\alpha_{rp}$ (equation 14).

$$\alpha_{rp} = \frac{k_{app,H}}{k_{app,L}} \tag{14}$$

where $k_{app,H}$ is the reaction rate for compound with the heavy isotope at the reaction position and $k_{app,L}$ is the overall degradation rate. The position-specific fractionation factor was derived from the microcosm isotope data using a procedure described in Elsner et al. (in preparation).

**Diffusion coefficients.** Two diffusion coefficients were differentiated and attributed to subspecies. The diffusion coefficient found from the literature was attributed to molecules with $^{12}C$ only which are dominant. Molecules containing one $^{13}C$ are assumed to diffuse slower because of its heavier weight. The corresponding diffusion coefficient ($D_H$) was calculated based on the diffusion coefficient of the molecule with $^{12}C$ only using (Cerling et al. 1991):

$$\frac{D_L}{D_H} = \left( \frac{M_{w1} + M_a}{M_{w1} M_a} \quad \frac{M_{w2} M_a}{M_{w2} + M_a} \right)^{1/2} \tag{15}$$

where $D_L$ and $D_H$ are the diffusion coefficients of $^{12}C$-compound and $^{13}C$-compound respectively, and where $M_{w1}$, $M_{w2}$ and $M_a$ are the atomic masses of $^{12}C$-compound, $^{13}C$-compound and average air, respectively.

**Table 2.** Parameter values taken for mathematical simulations of n-hexane

| | Carbon | | Hydrogen | |
|---|---|---|---|---|
| **Parameter** | **constant** | **decaying** | **constant** | **decaying** |
| $D_L$ ($m^2$/d) | 0.067 | 0.067 | 0.067 | 0.067 |
| $D_H$ | 0.0669 | 0.0669 | 0.0669 | 0.0669 |
| $K_{app,\,L}$ ($d^{-1}$) | 0.054 | 0.054 | 0.054 | 0.054 |
| $K_{app,\,H}$ | 0.0534 | 0.0534 | 0.0303 | 0.0303 |
| Source decay ($d^{-1}$) | --- | 0.049 | --- | 0.049 |
| $\varepsilon_{corr}$ (‰) | -10.43 | -10.43 | -439 | -439 |

$D_L$: from Werner et al. 2004
$K_{app,L}$: from Gaganis et al. 2004
Source decay: from Broholm et al. 2005

Having defined all the parameters, concentration profiles of subspecies I, II and III as a function of distance from the source were calculated using equation 7 that was implemented on Maple 9.0. Based on the concentration profiles of the subspecies, profiles for the total concentration and $\delta^{13}C$ of the compound were derived.

## Results and Discussion

### Microcosm experiments

After an acclimation period during the first series of tests (lag period before any degradation occurs), the 4-compound mixture was degraded with reaction rates in the following order: *n*-octane > *n*-hexane > aromatic compounds. Results yielded carbon isotope enrichment factors varying from -0.89 ‰ ± 0.09 for toluene to –2.89 ‰ ± 0.05 for benzene (Table 3.). In the cycloalkane experiment, methylcyclohexane was degraded preferentially over methylcyclopentane and both yielded small carbon isotope enrichment factors (Table 3.). In the *n*-alkane experiment, *n*-hexane was degraded preferentially over *n*-pentane and yielded a larger hydrogen enrichment factor of -132 ‰ compared of -97 ‰ for *n*-pentane.

**Table 3.** Carbon and hydrogen isotope enrichment factors obtained during aerobic biodegradation of alkanes,cycloalkanes and aromatic compounds.

| | Enrichment factors | |
|---|---|---|
| **Compounds** | **Hydrogen (‰)** | **Carbon (‰)** |
| n-Pentane (C5) | -97 ± 8 (n=2) | n.d. |
| n-Hexane (C6) | -132 ± 15 (n=2) | -2.58± 0.28 (n=3) |
| n-Octane (C8) | n.d. | -1.28 ± 0.11 (n=3) |
| Benzene (C6) | n.d. | -2.89 ± 0.05 (n=3) |
| Toluene (C7) | n.d. | -0.89 ± 0.09 (n=3) |
| Methylcyclopentane (C6) | n.d. | -1.53 ± 0.04 (n=2) |
| Methylcyclohexane (C7) | n.d. | -1.11 ± 0.11 (n=2) |
| 3-methylpentane (C6) | n.d. | -1.09 ± 0.03 (n=3) |

The carbon isotope enrichment factors obtained during aerobic degradation of benzene were similar to those obtained in other laboratory experiments with pure cultures in liquid media (Hunkeler et al. 2001) or in batch experiments with mixed cultures (Meckenstock and Richnow 2002).

Results for toluene are different from those of Sherwood Lollar et al. (1999) who found no isotope fractionation in microcosm experiments. On the other hand, Morasch et al. (2002) presented 2 different isotopic enrichment factors for the aerobic degradation of toluene depending on the enzyme that carries out the initial attack, being –3.3 ‰ for a methyl monoxygenase and –1.1 ‰ for ring monoxygenase. Comparison of the present results (toluene $\varepsilon = 0.89$ ‰) with that study would imply a dominating ring monoxygenase mechanism that degrades toluene in our soil microcosms. Our study also provides carbon isotope enrichment factors for short *n*-alkanes, and also for cycloalkanes and 3-methylpentane. Globally and within a compound class, carbon enrichment factors decrease with increasing number of carbons. This is related to the probability of having a $^{13}C$ at the reactive sites. Finally, hydrogen enrichment factors are found to be much larger than carbon enrichment factors, as known also from other studies (Mancini et al. 2003, Morasch et al. 2002, Hunkeler et al. 2001). In our study, the hydrogen enrichment factor for n-hexane degradation is 50 times larger than the carbon enrichment factor. This is mainly due to a larger relative mass difference in case of D/H compared to $^{13}C/^{12}C$; deuterium is 100 % heavier that hydrogen, compared to 8 % mass difference between $^{13}C$ and $^{12}C$.

## Field experiment

Figure 1 shows the evolution of $\delta^{13}C$ values forn-hexane in sampling points at a depth of 1m below the surface. Results indicate relatively constant $^{13}C/^{12}C$ ratios near the source over time when compare to those further away. The $\delta^{13}C$ values in the vicinity of the source are stable due to a constant supply by the source. At day 3 and to a lesser extent at day 6, decreasing $\delta^{13}C$ values with increasing distance from the source can be observed. These data strongly suggest that diffusion of light subspecies initially influence the isotope ratios. At day 22 and thereafter, more positive isotope ratios with respect to the source are measured, with an increasing shift towards more positive values over time. Finally, at day 114, all sampling points including those close to the source show elevated $\delta^{13}C$ values. This final increase is likely due to source depletion. Once the source is depleted, the shift towards positive values by biodegradation and diffusive loss of light molecules to the atmosphere is no longer balanced by supply of molecules with the original isotope signature at the source.

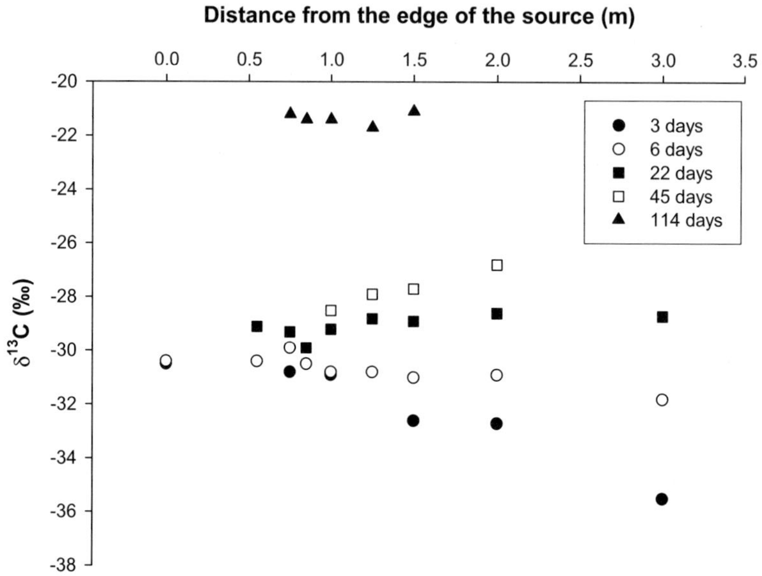


**Fig. 1.** Field experiment at Værløse, Denmark. Evolution of the $\delta^{13}C$ of *n*-hexane in the unsaturated zone over time and distance from the buried source made up of *n*-hexane mixed with 12 other compounds

Similar variations in shifts in $\delta^{13}C$ between source and sampling points, were also observed for other compounds while the magnitude of the shifts was variable. Table 4 summarizes the shifts observed for 7 selected compounds during the course of the field experiment. Among these compounds, methylcyclopentane and 3-methylpentane are those showing the greatest shifts, similar to the shifts of $\delta^{13}C$ of *n*-hexane. In previous laboratory batch and column experiments, these 3 compounds were found to be less degradable compare to the other compounds (Höhener et al. 2003), supporting here again a diffusion effect on the isotope ratio.

**Table 4.** Maximal shifts (Δ) of carbon isotopic ratios $\delta^{13}C$ in negative and positive direction observed during field experiment.

| **Compound** | **$\Delta\delta^{13}C$ (‰)** | |
|---|---|---|
| | **Max. shift in neg. direction** | **Max. shift in pos. direction** |
| Benzene | -2.6 | 7.7 |
| Toluene | -3.2 | 5.1 |
| n-octane | -2.8 | 5.7 |
| Methylcyclopentane | -5.7 | 8.0 |
| Methylcyclohexane | -3.2 | 4.5 |
| 3-Methylpentane | -4.1 | 10.6 |

n.d.: not determined

## Model simulations for *n*-hexane

To gain insight into the relative contribution of biodegradation and diffusion on the observed isotope evolution, three different scenarios were simulated. In the first scenario it was assumed that biodegradation only fractionates, in the second that diffusion only fractionates and in the third that both processes contribute to isotope fractionation. Note that even though only one process contributes to isotope fractionation in scenario 1 and 2, both processes are considered to be active.

Simulations for both carbon and hydrogen isotopes were performed with either a constant source of n-hexane (simulating a large spill) or a decaying source (simulating a small spill). It is important to note that we did not attempt to exactly reproduce the field data with the simulations but to get insights into the processes controlling the isotope ratios.

### *Isotope fractionation by biodegradation*

Figure 2 illustrates the evolution of the total compound concentration with distance from the edge of the source at different time steps. The constant source shows increasing concentration with time and approaches a steady state. In contrast, concentrations rapidly decrease of the decaying source scenario.

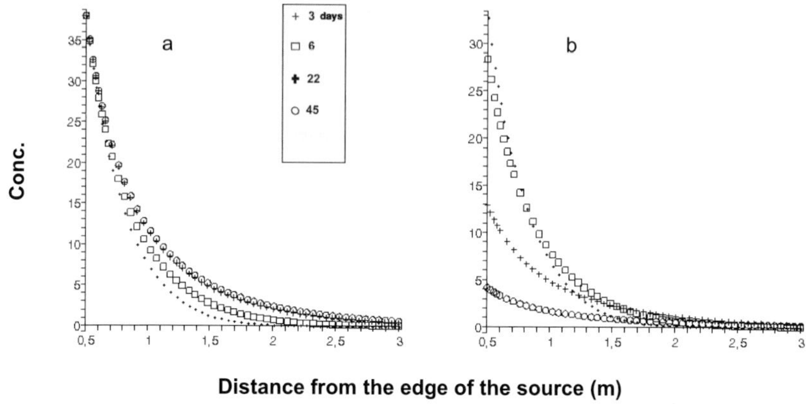


**Fig. 2. a,b.** Evolution of the concentration of carbon with time and distance for a constant (a) and a decaying source (b)

Figures 3 show simulation for $\delta^{13}C$ with either a constant source (a) and a decaying source (b) while figures 3 c and d show the corresponding simulations for δD. The *n*-hexane becomes at all times heavier with distance due to slower degradation of subspecies III. The shifts towards positive isotopic composition slightly increasing for carbon and considerably for hydrogen over time. As discussed above, the larger shift for hydrogen isotopes compared to carbon is due to the larger isotopic enrichment factor. The shifts in $\delta^{13}C$ or δD are larger for the decaying source than for a constant source. Even if the first order equation implies less biodegradation at lower concentration, the greater extend of biodegradation in the case of the decaying source creates larger shifts in $\delta^{13}C$ or δD.

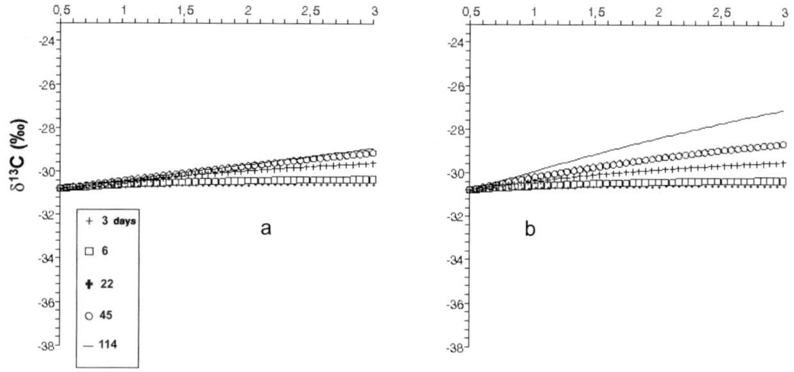

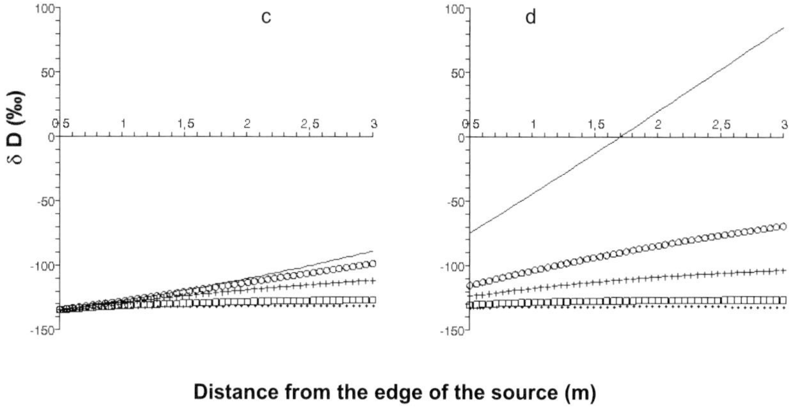


**Fig. 3. a,b,c,d.** Predicted values of $\delta^{13}C$ for *n*-hexane over time and distance when biodegradation controls the isotope fractionation, for a constant source (a) and a decaying source (b) and predicted values of δD for a constant source (c) and a decaying source (d)

## *Isotope fractionation by diffusion*

Figures 4 a, b, c and d show the predictions of carbon and hydrogen isotopic ratios of *n*-hexane under the assumption that only diffusion is contributing to isotope fractionation either for a constant and a decaying source. The results predict distinct negative shifts at early times (< 6 days) which are getting later less negative over time and distance for both atoms. The shift tends to level out with time because the molecules with $^{13}C$ catch up with time. However, even at day 114, shifts towards more negative values are still observed. Even though a shift is occurring for hydrogen, it would be less noticed as the shift stays in the limit of uncertainty of the measurement which is larger for hydrogen than for carbon. The different nature of the source (constant or decaying) has a small influence on the isotopic evolution. As the diffusive flux depends on the diffusion coefficient as well as the concentration gradient, smaller concentration gradient results in faster homogenization. Related to that, evolution of the isotopic ratio with a decaying source tends to level out slightly faster. The amplitudes of shift are equivalent for both carbon and hydrogen as in both cases, the mass of the molecule is increased by 1 by adding a $^{13}C$ or D atom.

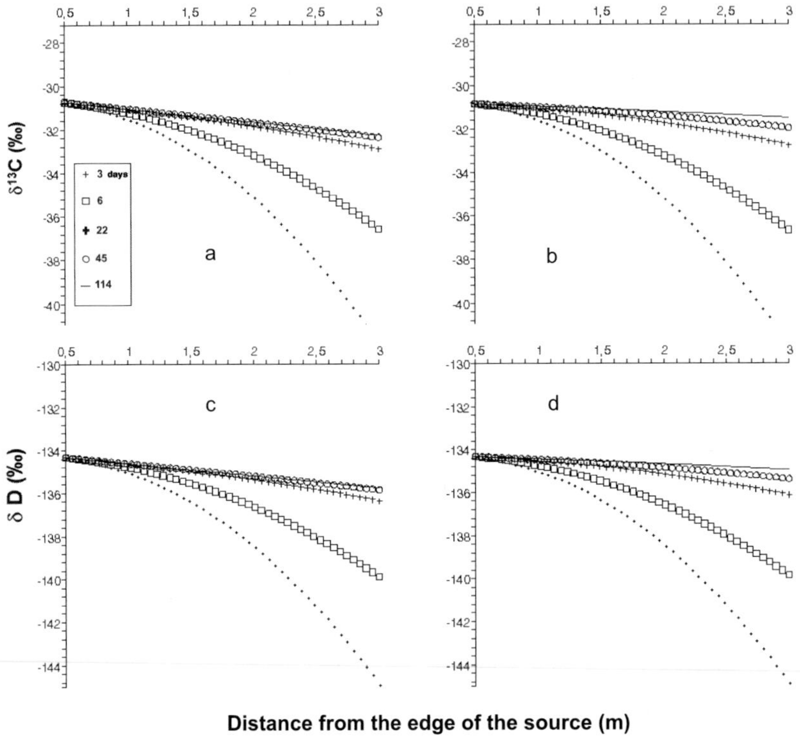


**Fig. 4. a,b,c,d.** Predicted values of $\delta^{13}C$ for *n*-hexane over time and distance when diffusion controls isotope fractionation, for a constant source (a) and a decaying source (b). Predicted values of δD for a constant source (c) and a decaying source (d)

## *Isotope fractionation by biodegradation and diffusion*

Figures 5 a and b and figures 5 c and d show simulation results assuming that both biodegradation and diffusion are responsible for the isotope fractionation of carbon and hydrogen isotopes with either a constant or a decaying source. These results reveal interesting conclusions for carbon and hydrogen isotopes. In case of carbon, the isotope evolution is similar to the one observed for diffusion as the only fractionating processes, especially at early times. Later (day 45, 114), the $\delta^{13}C$ increases to values more positive than the source, which is not observed when diffusion only fractionates. Qualitatively, the scenario with decaying source reproduces the field well except for day 114. In both simulation and field data, n-hexane is depleted in $^{13}C$ with distance at day 3 and 6, isotope ratios levels out by day 22 and become enriched in $^{13}C$ at day 45 with distance. The shift to $^{13}C$ enriched values for all points at 114 observed in the field but not in the simulations could be due to preferential loss of molecules with $^{12}C$ to the atmosphere. Loss to the atmosphere was not incorporated in the analytical model. The process will be incorporated in a numerical model in future studies. As showed in table 5, diffusion itself would explain around 95% of the carbon isotope shift occurring over the first 6 days, at 1.5 m away from the edge of the source for a decaying source. But starting from day 45, the biodegradation becomes responsible for more than 50 % of the isotope shift

**Table 5.** Calculated contribution of diffusion on the changing isotope ratio for carbon and hydrogen at 1.5 m from the edge of a constant and a decaying source

| | Proportion explained by diffusion (%) | | | |
|---|---|---|---|---|
| | Carbon | | Hydrogen | |
| **Day** | **C*** | **D*** | **C** | **D** |
| 3 | 95.6 | 95.7 | 54.5 | 55.4 |
| 6 | 85.2 | 95.7 | 26.7 | 26.7 |
| 22 | 55.6 | 85.7 | 6.7 | 3.5 |
| 45 | 42.1 | 47.7 | 4.1 | 1.0 |
| 114 | 40.0 | 28.6 | 3.5 | 0 |

* C: constant source, D: decaying source

In contrast to carbon, the obtained curve for hydrogen resemble more the simulation with biodegradation as only fractionating process shown above (Fig. 5 c and d). The larger influence of biodegradation on hydrogen isotope ratios can be explained by the fact that the addition of D to a molecules has a substantial influence on the biodegradation rate, while the effect on diffusion remains the same as for carbon. Consequently, the strong effect of biodegradation dominated over the effect of diffusion for both decaying and constant source. With a constant source, the effect of diffusion explains less than 4 % after 114 days. With a decaying source, the effect of diffusion is almost negligible already after 22 days (table 5).

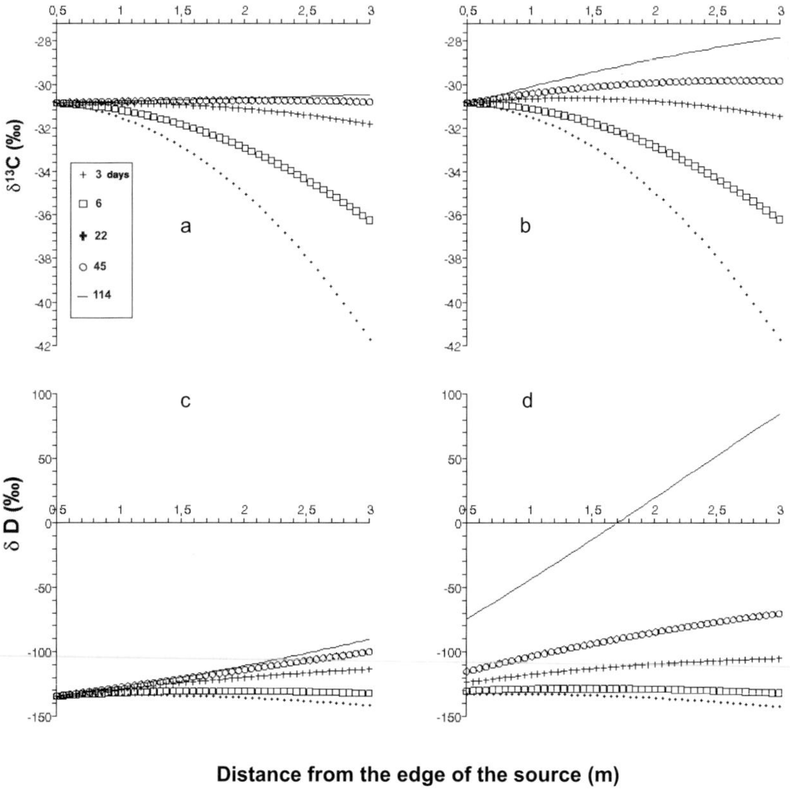


**Fig. 5. a,b,c,d.** Predicted values of $\delta^{13}C$ for *n*-hexane over time and distance when biodegradation and diffusion control the isotope fractionation, for a constant source (a) and a decaying source (b) and predicted values of δD for a constant source (c) and a decaying source (d)

## Conclusions

The laboratory experiments showed that carbon and hydrogen isotope fractionation occurs during aerobic degradation of short *n*-alkanes, cycloalkanes and aromatic compounds. The field experiment demonstrated that significant changes in $\delta^{13}C$ occur in the unsaturated zone over time and distance. Mathematical simulations of *n*-hexane diffusing through a homogeneous vadose zone made it possible to evaluate the relative importance of biodegradation and diffusion on the observed isotopic shifts of the contaminant. These findings underline the possibility of having isotope fractionation not caused by the biodegradation but by diffusion when investigating VOCs in the unsaturated zone. The isotope fractionation caused by diffusion is expected to be greater for small molecules. This complicates the use of isotope fractionation to assess the extent of biodegradation of a contaminant in the unsaturated zone, especially when the initial concentration and the age of the spill are unknown. Due to the higher isotope fractionation of hydrogen compared to carbon, hydrogen isotope ratio are likely better suited to trace biodegradation in the unsaturated zone at contaminated sites provided that compounds of interest show large hydrogen isotope fractionation. The larger hydrogen isotope fractionation quickly dominates over the effect of diffusion in the resulting $\delta D$. As the effect of diffusion tends toward uniformity in distance with time, carbon analysis should be used only for spills that are at least several months old.

## Acknowledgements

Funding was provided by the State Secretariat for Education and Research in frame of COST Action 629 and from Natural Science and Engineering Research Council, Canada. The authors thank Dr. Petros Gaganis, Dr. Richard Baltensperger and Prof Dr. Pierre Perrochet for useful hints concerning mathematical simulations

## References

Atlas RM (1981) Microbial degradation of petroleum hydrocarbons: an environmental perspective. Microbiological Reviews 45:180-209

Broholm M, Christophersen M, Maier U, Stenby E, Kjeldsen P (2005) Compositional evolution of the emplaced fuel source experiment in the vadose zone field experiment at Airbase Værløse, Denmark. Environ Sci Technol submitted

Cerling TE, Solomon DK, Quade J, Bowman JR (1991) On the isotopic composition of carbon in soil carbon dioxide. Geochim Cosmochim Acta 55:3403-3405

Christophersen M, Broholm M, Mosbaek H, Karapanagioti H, Kjeldsen P (2005) Emplaced fuel source experiment in the vadose zone at Airbase Vaerlose, Denmark: site characterization and hydrocarbon migration. J Contam Hydrol submitted

Clark I D, Fritz P (1997) Environmental isotopes in hydrogeology Lewis publisher, CRC Press LLC

Elsner M, Zwank L, Hunkeler D, Schwarzenbach RP (2005) A New Concept Linking Observable Stable Isotope Fractionation to Transformation Pathways of Organic Pollutants. Environ Sci Technol. In prep.

Gaganis P, Kjeldsen P, Burganos VN (2004) Modeling Natural Attenuation of Multicomponent Fuel Mixtures in the Vadose Zone: Use of Field Data and Evaluation of Biodegradation Effects. Vadose Zone J 3:1262-1275

Höhener P, Duwig C, Pasteris G, Kaufmann K, Dakhel N, Harms H (2003) Biodegradation of petroleum hydrocarbon vapours: laboratory studies on rates and kinetics in unsaturated alluvial sand. J Contam Hydrol 66:93-115

Hunkeler D, Andersen N, Aravana R, Bernasconi M, Butler BJ (2001) Hydrogen and carbon isotope fractionation during aerobic biodegradation of benzene. Environ Sci Technol 35:3462-3467

Kaufmann K, Christophersen M, Buttler A, Harms H, Höhener P (2004) Microbial community response to petroleum hydrocarbon contamination in the unsaturated zone at the experimental field site Værløse, Denmark. FEMS Microbiology Ecology 48-3:387-399

Kirtland BC, Aelion CM, Stone PA, (2005) Assessing in situ mineralization of recalcitrant organic compounds in vadose zone sediments using $\delta^{13}C$ and $^{14}C$ measurements. J Contam Hydrol 76:1-18

Mancini SA, Ulrich AC, Lacrampe-Couloume G, Sleep B, Edwards EA, Sherwood Lollar B (2003) Carbon and Hydrogen Isotopic Fractionation during Anaerobic Biodegradation of Benzene. Appl Environ Microbiol 69-1:191-198

Meckenstock RU, Morasch B, Griebler C, Richnow HH (2004) Stable isotope fractionation analysis as a tool to monitor biodegradation in contaminated aquifers. J Contam Hydrol 75:215-255

Meckenstock RU, Richnow HH (2002) Quantification of isotope fractionation in experiment with deuterium-labeled substrate. Appl Environ Microbiol 68:5206-5207

Morasch B, Richnow HH, Schink B, Vieth A, Meckenstock RU (2002) Carbon and hydrogen isotope fractionation during aerobic degradation of aromatic hydrocarbons. App Environ Microbiol 68:5191-5194

Pasteris G, Werner D, Kaufmann K, Höhener P (2002) Vapor phase transport and biodegradation of volatile fuel compounds in the non saturated zone: a large scale lysimeter experiement. Environ Sci Technol 36:30-39

Richnow HH, Annweiler E, Michaelis W, Meckenstock RU (2003) Microbial in situ degradation of aromatic hydrocarbons in a contaminated aquifer monitored by carbon isotope fractionation. J Contam Hydrol 65:101-120

Sherwood Lollar B, Slater GF, Ahad J, Sleep B, Spivack J, Brennan M, MacKenzie P (1999) Contrasting carbon isotope fractionation during biodegradation of trichloroethylene and toluene: Implications for intrinsic bioremediation. Org Geochem 30:813-820

Stehmeier LG, Francis M.McD, Jack TR, Diegor E, Winsor L, Abrajano Jr TA (1999) Field and in vitro evidence for in-situ bioremediation using compound specific $^{13}C/^{12}C$ ratio monitoring. Org Geochem 30:821-833

Steinbach A, Seifert R, Annweiler E, Michaelis W (2004) Hydrogen and carbon isotope fractionation during anaerobic biodegradation of aromatic hydrocarbons – A field study. Environ Sci Technol 38:609-616

Griebler C, Safinowski M, Vieth A, Richnow HH, Meckenstock RU (2004) Combined application of stable carbon isotope analysis and specific metabolites determination for assessing in situ degradation of aromatic hydrocarbons in a tar oil-contaminated aquifer. Environ Sci Technol 38:617-631

Werner D, Broholm M, Höhener P (2004) Simultaneous estimation of nonaqueous phase liquid (NAPL) saturation and diffusive fluxes of volatile organic compounds in the vadose zone. Ground Water Monitoring and Remediation, in press

Werner D, Höhener P (2003) In situ method to measure effective and sorption-affected gas-phase diffusion coefficients in soils. Environ Sci Technol 37-11:2502-2510

# A model assessing bioavailability of persistent organic pollutants in soil

G. Fragoulis, M. Trevisan, E. Puglisi, E. Capri

Università Cattolica del Sacro Cuore, Istituto di Chimica Agraria ed Ambientale, sezione Chimica Vegetale. Via Emilia Parmense 84, 29100 Piacenza, Italy. Tel:+39 0523 599345, fax:+39 0523 599217, e-mail: george:fragoulis@unicatt.it

## Abstract

In this paper a model is proposed for describing persistent organic pollutants (POPs) bioavailability in soil. The model is written in Fortran 90 and describes POPs' behaviour as resulting from four different processes: sorption-desorption equilibrium, slow diffusion (aged fraction), fast irreversible sorption (bound residues) and biodegradation of the bioavailable fraction.

The POP sorption to soil surfaces is described assuming a rapid rate of sorption-desorption to and from soil surfaces and a slower rate of diffusion into the internal matrix (aging). Biodegradation is described as resulting from bacterial growth using sigmoidal Monod kinetics for the contaminant dissolved in soil solution (for non-hydrophobic compounds) and first order kinetics for the degradation of the sorbed-available fraction. In the case of hydrophobic compounds, first-order kinetics is employed to describe also the degradation from the soil solution. Sorption and diffusion are approximated by first order kinetics. Finally, the formation of bound residues is described using an exponential saturation equation. The rate constants for the different processes are estimated using linear and non lincar first order kinetics approaches. The rate constants of the irreversible processes are estimated from experimental data. Model evaluation was performed using data from previous experiments with phenanthrene as test compound.

## Keywords

Modelling; Bioavailability

## Introduction

Persistent organic pollutants (POPs), because of their high toxicity, long environmental persistence and high bioaccumulation factors, are considered serious pollutants. In the last twenty years their potential impact to the environment has been analysed and described extensively.

The fate and behaviour of organic pollutants in soil is governed by many factors, including soil characteristics, compound properties and environmental factors (Reid et al. 2000a, Chung and Alexander 2002, Nam et al. 2003). Polycyclic Aromatic Hydrocarbons (PAHs) represent one of the most widespread organic pollutants in the environment. PAHs are formed as by-products of incomplete combustion of organic materials and are emitted in the environment mainly from anthropogenic sources. A great environmental concern is associated with their presence, as some of them were demonstrated to be carcinogenic. They are highly hydrophobic, recalcitrant to degradation and due to their affinity for fatty tissues they tend to be bioaccumulated within food-chains.

Organic pollutants have traditionally been studied with exhaustive extraction techniques and these results are often coupled to risk assessment models to add a biological interpretation. However, this approach does not take into account two important processes, bioavailability and aging (Hatzinger et al. 1995, Laor et al. 1996). Bioavailability is a dynamic rather than a static process. Time-dependent losses in bioavailability have been demonstrated for several organic and inorganic chemicals. Upon entering a soil system, pollutants first contaminate the macropores and the particle surfaces containing relatively few bacteria. They then diffuse into smaller pores where biotransformation may take place, given that the environmental conditions are in favour of microbial activity. Pollutants also diffuse into extremely small pores of the mineral or organic soil fraction. The result of the slow entrapment of chemicals into sites within the soil matrix that are not accessible to the microbial biomass is called aging (Alexander 2000). As POPs persist within a soil or sediment matrix, they become increasingly resistant to desorption and thus less bioavailable (Gevao et al. 2000). The nature and extent of aging processes depends on soil and compound properties as a result of physical, chemical and biological interactions (Chung and Alexander 1998, Northcott et al. 2001). Pollutants become unavailable when the rate mass transfer is zero (i.e. the case of aged and bound residues).

Reliable and representative techniques for the assessment of the bioavailable fraction of organic contaminants are still lacking. The problem is complicated by the fact that this fraction varies depending on the physico-chemical properties of the molecule; for the same contaminant the type of soil and target organisms have also to be considered (Chung and Alexander 1998).

Since only very recently the concept of bioavailability has been introduced in the literature and the mechanisms of aging and bound residues formation are not yet fully understood, few bioavailability models exist by now. Scow and Hutson (1992) presented the DSB-Model using terms for linear sorption to particulate surfaces, diffusion within soil aggregates and biodegradation using Monod kinetics. Shelton and Doherty (1997) proposed a conceptual model that describes pesticide biodegradation rates, accounting for bioavailability and microbial growth using Monod kinetics. Cornelissen et al. (1998) used a first-order, two-compartment model to describe desorption. Altfelder et al. (2000) described the role of slow kinetics in non-singular sorption of organic compounds in soil, using a non-linear two-stage sorption model. A more sophisticated model coupling intraparticle diffusion/sorption, non-linear sorption, and biodegradation was suggested by Karapanagioti et al. (2001). However, the focus of the latter model is on predicting the magnitude of natural attenuation of solutes undergoing transport and mobile–immobile domain mass transfer in porous media, rather on describing the patterns of bioavailability.

The model proposed in this paper is a conceptual five compartment one, which describes the soluble, sorbed (bioavailable), aged (non bioavailable), bound (non bioavailable) and degraded fraction of a contaminant entering a soil system. This model can be used along with innovative non-exhaustive extraction techniques, such as cyclodextrins and hydrophobic resins, in order to assess the bioavailability of hydrophobic organic compounds in soils.

## Processes described

### Adsorption-desorption-aging kinetics

The sorption is composed of two distinct processes: a rapid rate of adsorption to soil surfaces, which is in rapid equilibrium with the soluble fraction and remains readily bioavailable, and a slower rate of diffusion into soil aggregates-organic matter particles, which is not readily available for degradation. The latter is the aged fraction of the pollutant and is described with the first order rate constant $k_2$. The rapid rate of sorption-desorption to and from soil surfaces is described by first order kinetics. The ratio of the pollutant concentration sorbed to soil surfaces to pollutant in soil solution is estimated by the equation: $K_d$=om $K_{oc}$. An initial estimation is required for the first order rate constant for sorption to soil surface ($k_1$); given this estimate, the first order rate constant for desorption from the soil ($k_3$) can be obtained from the equation: $K_d= k_1/ k_3$. If known, the $K_{oc}$ is used directly to estimate the $K_d$. In the case that $K_{oc}$ is not known, the value of the $K_{ow}$ is used to estimate it, by solving the following equation: $\log K_{oc}= 0.92\log(K_{ow})-0.23$.

### Bound Residues

When the rate of mass transfer to the degrading microorganisms is zero, pollutants form bound residues, thus becoming unavailable. The formation of bound residues is described using an exponential saturation equation: $bound=boundF(1-e^{-k_4t})$ where $k_4$ is the rate constant for formation of bound residues and boundF is the highest relative concentration that bound residues can reach and it is proportional to the initial sorbed concentration.

### Biodegradation

Rates of contaminants degradation have traditionally been described using first order kinetics where the rate is assumed to be dependent only on the pesticide concentration. However, in cases where contaminants are used as growth substrates by soil microorganisms, accelerated biodegradation is observed that is best described by sigmoidal kinetics. When growth linked processes are observed (bacterial growth), Monod kinetics is more appropriate; on the other hand Michaelis-Menten kinetics is mostly used to describe non-growth linked biodegradation. In this model only in the case of non-hydrophobic substances Monod kinetics is used to describe the degradation from the substrates in soil solution. In the case of sorption dependent degradation the biodegradation can occur also from the sorbed part of the substrate (as in the case of phenanthrene), and first order kinetics are used to describe the degradation from the sorbed and the soluble phase. The first order degradation rate constant is estimated from the equation: $k_t$=ln(2)/DT50, where DT50 is the pollutant's half life in days. In figure 1 is given the box diagram and the equations that govern the conceptual model.

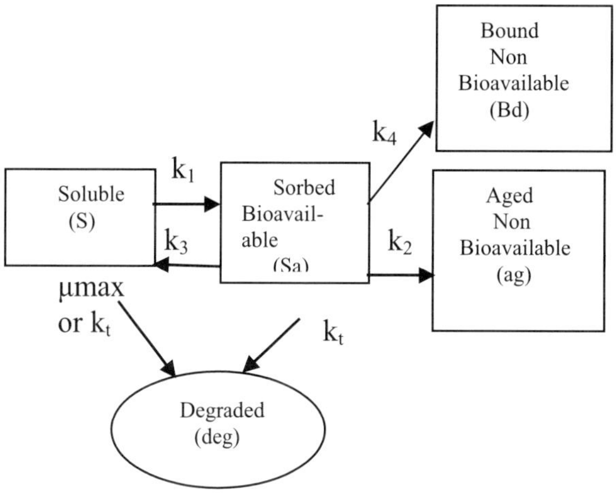


$$\frac{\partial S}{\partial t} = -k_1 S + k_3 Sa - \frac{\mu maxSB}{K_s + S} .or \ \frac{\partial S}{\partial t} = -(k_1 + k_t)S + k_3 Sa$$

$$\frac{\partial Sa}{\partial t} = k_1 S - (k_3 + k_2 + k_t + k_4) Sa$$

$$\frac{\partial ag}{\partial t} = k_2 Sa \quad Bd = boundF\left(1 - e^{-k_4 t}\right)$$

$$\frac{\partial \deg}{\partial t} = k_t Sa + \frac{\mu \max SB}{K_s + S} .or \ \frac{\partial \deg}{\partial t} = k_t (Sa + S)$$

**Fig. 1.** Box diagram and model equations (∂S/∂t=time variation of substrate concentration in soil solution, ∂Sa/∂t= time variation of substrate sorbed concentration, ∂ag/∂t= time variation of substrate aged concentration, ∂Bd/∂t= time variation of substrate bound concentration, ∂deg/∂t= time variation of substrate degraded concentration, µmax=maximum contaminant utilization/mass of microorganisms (1/h), B=total microbial concentration (mg/Kg), Ks=contaminant half-saturation constant (mg/L), $k_1$=first order rate constant for sorption to soil om (1/day), $k_2$=first order rate constant for aging (1/day), $k_3$=first order rate constant for desorption from the soil om (1/day), $k_4$=first order rate constant for bound residues (1/day) $k_t$=first order rate constant for degradation, boundF=maximum allowed bound residues concentration).

## Materials and Methods

### Model compound

In the present study phenanthrene ($C_4H_{10}$), a neutral organic compound that contains three aromatic rings, was selected as a representative PAH. It has environmental properties such as aqueous solubility, octanol–water partition coefficient ($K_{ow}$), and vapor pressure that are similar to other PAHs, such as acenaphthene, fluoranthene, and fluorene. The parent structure of the phenanthrene is similar to the structure of higher molecular weight and more carcinogenic PAHs, such as benzo(a)pyrene, so it is hypothesised that phenanthrene is an adequate representative compound that would give a general indication of PAH behaviour (Khodadoust et al. 2004). All PAHs are hydrophobic, and phenanthrene has an aqueous solubility of 1.1 mg/L at 25°C.

### Chemical assessment of the bioavailable fraction

The bioavailable fraction of chemicals has usually been associated with the quantity which is easily desorbed; in turn such quantity is determined by extraction of the aqueous phase. With hydrophobic organic compounds such as PAHs, poorly soluble and with a strong affinity towards the organic matter, this method has great limitations, since very low amounts of these compounds will be in the soil solution, while a large labile and potentially bioavailable pool remains on the solid phase. To address this problem some authors (Wang and Brusseau 1993, Reid et al. 2000b, Puglisi et al. 2003) have proposed to enhance the solubility of this contaminants with cyclodextrins solutions. Cyclodextrins are cyclic oligomers of a-D-glucose formed by the action of enzymes on starch. They present a hydrophilic surface and an internal hydrophobic cavity. Although fairly soluble in water, they are at the same time able to entrap hydrophobic molecules inside their cavity, enhancing their solubility. When internal cavity is substituted with functional groups, such as in hydroxypropyl-β-cyclodextrins (HPCD) the solubilization effect towards hydrophobic compounds is furthered. Cyclodextrine extractions follow an equilibrium kinetics and the quantity extracted is strongly correlated to the bioavailable fraction (Reid et al. 2000b).

## Experimental Section

### *Soil Spiking*

Soil samples were collected in South Italy, in the Province of Bari (località Terra Ripagnola). Soils in this area are loamy, 25 cm deep, with a red colour due to high content of iron oxides, hence the denomination "red soils". Soil samples were passed trough a 2 mm sieve to remove roots and other vascular materials. Samples were spiked with a dilution mixing method with acetone as a carrier solvent and then divided in separate microcosms of 20 g each. Sterilisation was then carried out on treatment 1 and 2 by means of γ-irradiation (5 Mrads). The experiment was carried out on both sterilised and non sterilised samples, and 3 treatments (5 replicates each) were compared:

1. sterilised soil contaminated with 10 ppm of phenanthrene
2. sterilised soil contaminated with 100 ppm of phenanthrene
3. non sterilised soil contaminated with 100 ppm of phenanthrene

### *Phenanthrene extractions from soil*

HPCD and methanol extractions were carried out in sterilised and non sterilised soil in different time-steps. The quantity extracted with HPCD represents the bioavailable fraction of the contaminant. The quantity extracted with methanol represents the sum of the bioavailable,and the aged fraction, while the total initial compound concentration minus the quantity extracted with methanol (bioavailable+aged) at the sterilised soil gives the bound fraction. Finally, the total initial compound concentration minus the quantities extracted with methanol, minus the bound fraction at the non-sterilised soil, gives the biodegraded fraction of the chemical at each time-step.

Regarding the HPCD extraction, samples of 1 g of soil were weighted in centrifuge policarbonate tubes and added with 20 mL of HPCD 50 mM. The tubes were sealed, placed on an orbital shaker at 180 revertants/min for 12 hours and then centrifuged at 4000 revertants/min for 1 hour. Then the supernatant was re-centrifuged for 10min, filtered and finally injected in HPLC for phenanthrene quantification.

Regarding the methanol extraction, samples of 1 g of soil were weighted in centrifuge policarbonate tubes and added with 20 mL of methanol. The tubes were sealed, placed on an orbital shaker at 180 revertants/min for 12 hours and then centrifuged at 4000 revertants/min for 1 hour. After centrifugation the supernatant was filtered and injected in HPLC for phenanthrene quantification.

## Results

In the case of phenanthrene, because of its hydrophobicity, the biodegradation can occur not only from the concentration in the soil solution but also from the sorbed fraction of the substrate (sorbed/bioavailable). For this reason first order kinetics are used to describe the degradation from the sorbed and the soluble phase. Since in the experimental results the extraction with HPCD represents the bioavailable fraction of the phenanthrene (soluble+sorbed/bioavailable) the first order degradation constant $k_t$ for the soluble and the sorbed fraction is the same. The first order rate constant for bound residues, $k_4$, was estimated from the experiment with the sterilised soil. The experimental results, with the non-sterilised soil, showed that the degradation of the phenanthrene started after 50 days of the contamination. On this account a lag-phase time of 50 days was predicted before biodegradation to start. The half-life of the phenanthrene was estimated in 30 days.

**Table 1.** Sorption-desorption, diffusion, degradation and bound residues formation kinetic constants

| initial concentration = 10 and 100 | | {mg/kg} |
|---|---|---|
| dt50 = 30 | | {days} |
| tlag = 50 | | {days} |
| kt = 0 | ig 0 < t < tlag | |
| kt = ln(2)/dt50 | if t > tlag | {l/day} |
| logKow = 4.2 | | |
| log(Koc) = 0.92*log(Kow) – 0.23 | | |
| om = 0.01 | | {kg/kg} |
| Kd = om*Koc | | |
| k1 = 0.02 | | {l/day} |
| k2 = 0.003 | | {l/day} |
| k3 = k1/kd | | {l/day} |
| k4 = 0.1 | | {l/day} |
| boundf = 15 % initial concentration | | {mg/kg} |

Table 1 summarises the first order rate constants for sorption-desorption, degradation, aging and bound residues formation for the non-sterilised soil.

Figure 2 shows the predicted concentrations, as a percentage of the initial applied concentration, of bioavailable (soluble+sorbed), non bioavailable (aged+bound) and degraded fractions of phenanthrene in the non-sterilised soil. Figure 3 compares modelled and measured concentrations of phenanthrene in the non-sterilised soil. The model was run with initial phenanthrene concentrations of 10 and 100 ppm with similar results. The model accounts for the water solubility of the substrate and does not allow the concentration in the soil solution to exceed it.

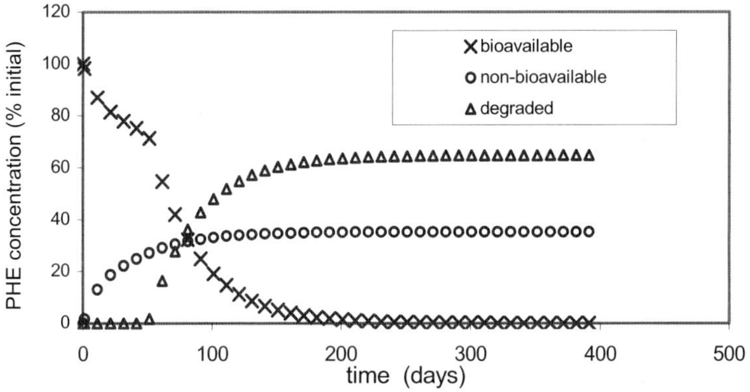


**Fig. 2.** Model predictions for bioavailable, non bioavailable and degraded phenanthrene

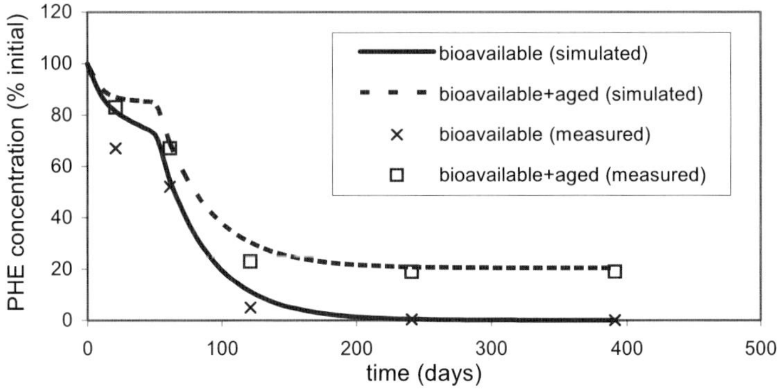


**Fig. 3** Comparison between model predictions and measured concentrations of phenanthrene.

## Conclusions

The proposed five compartment model can be used along with experimental data to predict the organic contaminants bioavailability in the course of time. Non-exhaustive extraction techniques, for bioavailability assessment of hydrophobic organic compounds in soils, such as cyclodextrins and hydrophobic resins, can provide the necessary data to calibrate and evaluate the model. The bioavailable fraction of the substance is the fraction in the soil solution and, in the case of hydrophobic organic compounds, also the fraction sorbed to the soil surfaces. First order kinetics are used to describe the degradation of the sorbed fraction, while sigmoidal Monod kinetics describe the degradation of the substrate in soil solution, if data are available. First order kinetics are used for the slow diffusion into soil aggregates-organic matter particles (aged fraction) and an exponential equation describes the formation of bound non-available residues. The model, after being calibrated using experimental data, described satisfyingly the phenanthrene sorption-desorption, aging, degradation and formation of bound residues for the range of 10-100ppm of initial phenanthrene concentration.

## References

Alexander M (2000) Bioavailability and overestimation of risk from environmental pollutants. Environ Sci Technol 34:4259-4265

Altfelder S, Streck T, Richter J (2000) Non-singular sorption of organic compounds in soil: the role of slow kinetics. J Environ Qual 29:917-925

Chung N, Alexander M (1998) Differences in sequestration and bioavailability of organic compounds aged in dissimilar soils. Environ Sci Technol 32:855-860

Chung N, Alexander M (2002) Effect of soil properties on bioavailability and extractability of phenanthrene and atrazine sequestered in soil. Chemosphere 48:109-115

Cornelissen G, Rigterink H, Ferdinandy MMA, Van Noort PCM (1998) Rapidly desorbing fractions of PAHs in contaminated sediments as a predictor of the extend of bioremediation. Environ Sci Technol 32(7):966–970

Gevao B, Semple KT, Jones KC (2000) Bound pesticide residues in soil: a review. Environmental Pollution 108:3-14

Hatzinger PB, Alexander M (1995) Effect of aging of chemicals in soil on their biodegradability and extractability. Environ Sci Technol 29:537-545

Karapanagioti HK, Gossard CM, Strevett KA, Kolar RL, Sabatini DA (2001) Model coupling intraparticle diffusionrsorption, nonlinear sorption, and biodegradation processes. Journal of Contaminant Hydrology 48:1–21

Khodadoust AP, Krishna RR, Kranti Maturi (2004) Removal of Nickel and Phenanthrene from Kaolin Soil Using Different Extractants. Environmental Engineering Science 21(6):691-704

Laor Y, Strom PF, Farmer WJ (1996) The effect of sorption on phenanthrene bioavailability. Journal of Biotechnology 51:227-234

Nam K, Kim JY, Oh DI (2003) Effect of soil aggregation on the biodegradation of phenanthrene aged in soil. Environmental Pollution 121:147-151

Northcott GL, Joses K (2001) Partitioning, extractability and formation of nonextractable PAH residues in soil.1. Compound differences in aging and sequestration. Environ Sci Technol 35:1103-1110.

Puglisi E, Patterson CJ, Paton GI (2003) Non-exhaustive extraction techniques (NEETs) for bioavailability assessment of organic hydrophobic compounds in soils. Agronomie 23:755-756

Reid BJ, Jones KC, Semple KT (2000a) Bioavailability of persistent organic pollutants in soils and sediments-a perspective on mechanisms, consequences and assessment. Environmental Pollution 108:103-112

Reid BJ, Stokes JD, Jones KC, Semple KT (2000b) Non-exhaustive cyclodextrine-based extraction technique for the evaluation of PAH bioavailability (2000b). Environ Sci Technol 34:3174-3179

Scow KM, Hutson J (1992) Effect of diffusion and sorption on the kinetics of biodegradation: Theoretical considerations. Soil Sci Soc Am J 56:119-127

Shelton DR, Doherty MA (1997) A model describing pesticide bioavailability and biodegradation in soil. Soil Sci Soc Am J 61:1078-1084

Wang X, Brusseau ML (1993) Solubilization of some low-polarity organic compounds by hydroxypropyl-cyclodextrin. Environ Sci Technol 27:2821–2825

# Geochemical changes under variably saturated conditions during artificial recharge via ponded infiltration – A field study

J. Greskowiak[1], G. Massmann[2], H. Prommer[3], G. Nützmann[1], A. Pekdeger[2]

[1] Leibniz-Institut of Freshwater Ecology and Inland Fisheries, Germany
[2] Free University of Berlin, Germany
[3] Utrecht University, The Netherlands

## Abstract

Artificial groundwater recharge by percolation through the unsaturated zone is an important technique to enhance the water quality for drinking water supplies. Purification of the infiltration water results mainly from microbially mediated redox-reactions that involve the degradation of a wide range of organic substances. The aim of this study was to identify the spatial and temporal distribution of the redox zones that develop beneath an artificial recharge pond, a system that is characterised by regular hydraulic changes between the saturated and unsaturated condition within every operational cycle. The most significant hydraulic changes result from the formation and removal of a clogging layer at the pond's bottom. Geochemical analyses of suction cup water as well as oxygen and hydraulic measurements showed that generally nitrate and manganese reducing conditions dominated below the pond as long as water saturated conditions prevailed. Iron and sulphate reduction occurred only in patchily distributed zones directly below the clogging layer and resulted from chemical and physical heterogeneity. When the sediment below the clogging layer became unsaturated, atmospheric oxygen penetrated from the pond fringes into this region, allowing re-oxidation of previously formed sulphide minerals.

## Keywords

artificial recharge; saturated/unsaturated zone; Terminal Electron Acceptor Processes (TEAPs); biodegradation; redox

## Introduction

Artificial recharge of groundwater by techniques such as riverbank filtration (Ray et al. 2002), aquifer storage and recovery (ASR, e.g., Herczeg et al. 2004), deep well injection (e.g., Stuyfzand 1998, Prommer and Stuyfzand 2005) and infiltration ponds (e.g., Bouwer 2002) are increasingly important for augmentation of groundwater resources and drinking water supplies. Infiltration ponds are commonly used either to purify partially treated sewage effluent or to enhance the quality of surface water (Bouwer 1991, Asano 1992). The enhancement of surface water quality is achieved by its percolation through an unsaturated soil zone, where a number of physical (filtration), chemical (e.g., adsorption) and biological processes (biodegradation) purification processes occur. Of these, biodegradation is generally considered to be the most important process in this context (Quanrud et al. 1996). The oxidation of degradable organic substances is linked to the reduction of terminal electron acceptors (TEAs) such as oxygen, nitrate, Mn- and Fe-oxides, and sulphate, which are typically consumed in a sequential order (Champ et al. 1979). The partial or complete mineralisation of organic carbons and the resulting consumption of TEAs can trigger a variety of additional geochemical reactions due to the release or consumption of protons and other reactants (Eckert and Appelo 2002) and can therefore alter both the water quality and the composition of the aquifer matrix.

Numerous investigations on the fate of a wide range of organic substances such as trace organic compounds, algal toxins and pathogens have been undertaken during infiltration of surface water (e.g., Fujita et al. 1998, Lindroos et al. 2002, Grützmacher et al. 2002, Långmark et al. 2004). In contrast, studies that simultaneously investigate the inorganic chemistry and which characterise the spatial distribution and dynamics of redox environments in ponded infiltration systems are rather scarce (e.g., Johnson et al. 1999, Brun and Broholm 2001). As local redox conditions affect the fate of various organic pollutants such as pharmaceutically active compounds (e.g., Holm et al. 1995), pesticides (e.g., Agertved et al. 1992) or halogenated compounds (e.g., Bosma at al. 1996), it is important to understand the linkages between prevailing hydraulic regimes and the resulting geochemical conditions.

The present paper summarises the research carried out for an artificial recharge pond near Lake Tegel (Berlin, Germany), where surface water is infiltrated in a cyclic manner. The main objectives of this study were

- to identify the transient unsaturated/saturated water regime during an operational recharge cycle, and
- to map the spatial and temporal distribution of the redox environment that developed under such a transient hydraulic regime.

Full details of the study can be found in Greskowiak et al. (2005).

## Background

### Pond operation

The study site is one of three recharge ponds located in the centre of a triangular shaped production well field near Lake Tegel in the northwest of Berlin, Germany (Figure 1). The pond's bottom is located 3 m below the ground surface. After filtration through a microstrainer, the surface water extracted from Lake Tegel is discharged into the pond. The recharge rate at the beginning of an operational cycle is typically 3 m/d and decreases to values as low as 0.3 m/d as a result of clogging processes occurring at the bottom of the pond. Clogging is caused by physical, chemical and biological effects and is known to reduce the infiltration rate in recharge basins to as little as one-tenth of the original infiltration rate (Bouwer 1978). Whenever the infiltration rate becomes too low, the operator of the ponds (Berliner Wasserbetriebe) removes the clogging layer, which restores the original hydraulic conductivity. After that, a new, typically 3-4 month-long operational cycle recommences.

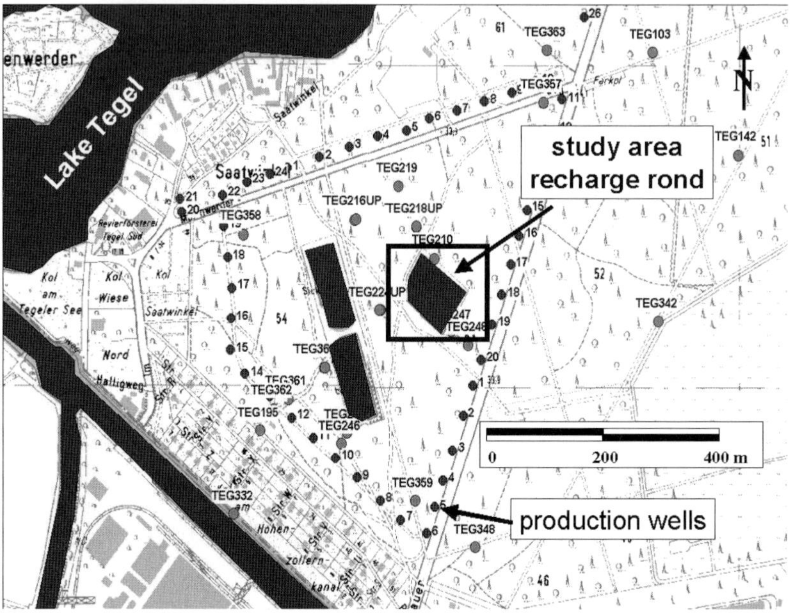


**Fig. 1.** Study area.

## Hydrogeology

The Quaternary sediment beneath the recharge pond extends to approximately 15 m below the ground surface. It consists mainly of fluvial medium sized sand deposits.

The hydraulic conductivity of the aquifer sediment is generally about $10^{-4}$-$10^{-3}$ m/s but can occasionally be as low as $10^{-5}$ m/s, in particular at depth between 0 m and 2 m below the bottom of the pond, where silt is sporadically present (Pekdeger et al. 2002). Reducible forms of iron and manganese minerals are in concentrations of up to 1.2 g/kg and 0.1 g/kg respectively. The sediment's inorganic carbon concentration is about 0.1-2 g/kg, which suggests the presence of calcite. The organic carbon content of the sediment generally varies across two orders of magnitude from 0.2 g/kg to 20 g/kg, except for a few thin coal layers that contained organic carbon concentrations of about 150 g/kg. Total sulphur concentrations range from 0.1 g/kg to 2.3 g/kg. A more detailed description of the hydraulic properties and the mineral composition of the sediments can be found in Pekdeger et al. (2002).

The water composition in the pond is generally characterised by relatively high sulphate concentrations of about 140 mg/l. The pH is about 8 - 8.5 and bicarbonate concentrations range from 150 to 190 mg/l. With calcium concentrations typically around 80 - 90 mg/l, the surface water is generally oversaturated with respect to calcite ($SI_{Calcite} \approx 0.7$). Nitrate concentrations can vary from 0 to 10 mg/l. The concentration of dissolved organic carbon (DOC) is rather low and ranges from 5 to 11 mg/l. Concentrations of dissolved iron and manganese are typically below the detection limit.

The decrease of the recharge rate leads to a decline of the groundwater below the pond. As a result, the porous medium beneath the pond becomes unsaturated. This shift occurs every operational cycle.

## Materials and Methods

From July 2003 until December 2003, sampling of (i) the pond water, (ii) groundwater from an observation well (screened at 6-8 m below the pond), and (iii) water extracted from four ceramic suctions cups at 50 cm, 100 cm, 150 cm and 200 cm below the pond was carried out on a weekly base. Water content and pressure heads were continuously recorded at two different depths, i.e., 50 cm and 150 cm below the pond. The locations of all these probes are shown schematically in Figure 2. Alkalinity and pH were measured in the field immediately after sampling. Alkalinity was determined by Gran Titration. All samples were analysed for major ions and dissolved organic carbon (DOC) using standard methods. Data from samples with a charge balance error of more than 5 % were discarded. Dissolved oxygen (DO) concentrations were measured by optical sensor-type oxygen probes (Hecht and Kölling 2001) placed next to the suction cups.

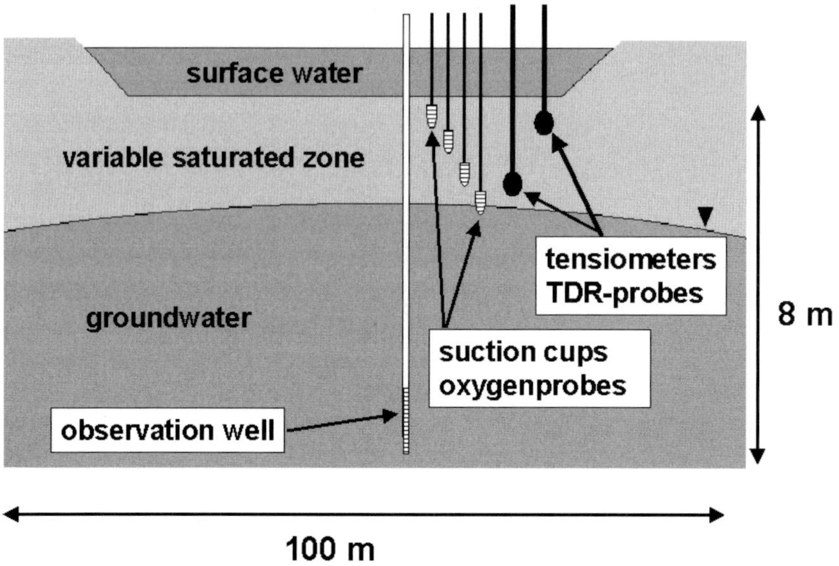


**Fig. 2.** Schematic cross-section of study area and locations of sample collection and measurement devices.

## Results and discussion

### Hydraulic conditions

Measured time-series of water contents (Figure 3) and pressure heads (data not shown) below the pond indicated that the transient hydraulic conditions of each operational cycle could be classified hydraulically into four different major stages. Stage 1 marked the transition from unsaturated to saturated conditions beneath the pond (Figure 3), which occurred soon after the removal of the clogging layer once infiltration was restarted. The saturated conditions prevailed for approximately 50 days (Stage 2) until the of the groundwater table dropped abruptly within a few days (data not shown). The decline of the water table forced air from the pond margins to the centre beneath the recharge pond and water contents fell from about 40-45 % to 30-35 % (Figure 3). The unsaturated conditions prevailed for 40 days (Stage 3). During the last period of the operational cycle (Stage 4) the pond was empty and the redevelopment of the pond took place. Hence, no recharge occurred. A schematic diagram of the relevant hydraulic stages is given in Figure 4.

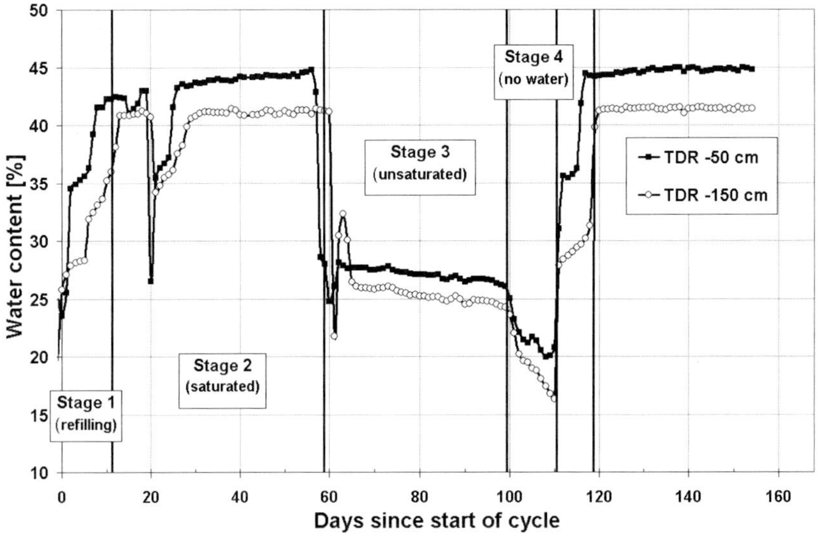


**Fig. 3.** Water contents in 50 cm and 150 cm depth beneath the pond.

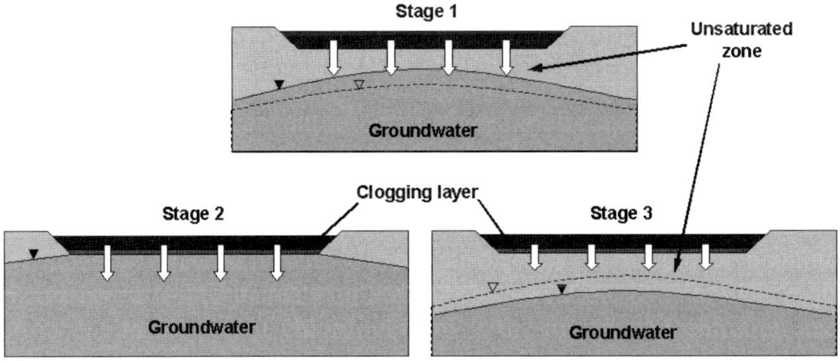


**Fig. 4.** Schematic cross-sectional diagram of the pond for the Stages 1, 2 and 3.

## Dynamics and Distribution of redox environments

During Stage 1, oxygen concentrations declined from about 5 mg/l to zero at all observed depths (Figure 5), i.e., oxygen was consumed within the first 50 cm below the pond. During the entire Stage 2, DO concentrations remained zero at all observed depths. About 20 to 30 days after the beginning of Stage 2, nitrate concentrations at various depths started to drop below the concentrations found in the surface water (Figure 6). At 50 cm and 100 cm depth nitrate was totally depleted within a few days, whereas the concentrations in the groundwater decreased more slowly (Figure 6). In contrast, at 150 cm and 200 cm depth, nitrate was not fully consumed.

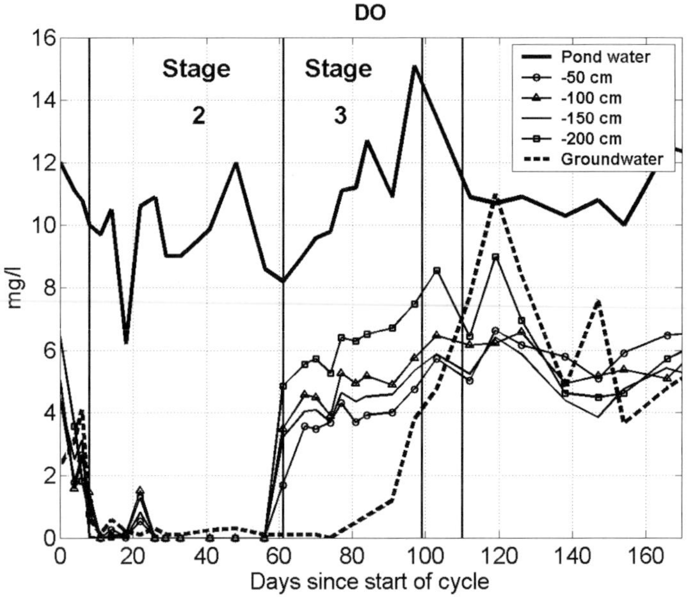


**Fig. 5.** Oxygen concentrations in the pond and groundwater and at depths of 50 cm, 100 cm, 150 cm and 200 cm beneath the pond.

At locations where nitrate was depleted, the redox-environment shifted to manganese reducing conditions, as indicated by an increase of manganese ($Mn^{2+}$) concentrations (up to 8 mg/l, see Figure 6). At a depth of 100 cm below the pond, where $Mn^{2+}$ concentrations were highest, iron ($Fe^{2+}$) concentrations increased to 0.3 mg/l (data not shown), indicating that iron reduction took place during Stage 2. The heightened $Mn^{2+}$ concentrations at these depths lasted for approximately 10 days before they started to decline again. The decline was most likely the result of a decrease in water temperature (data not shown), associated with a decrease of microbial activity and coincided with a simultaneous rise of nitrate concentrations beneath the pond (Figure 6). During Stage 1 and Stage 2, sulphate concentrations were generally around 140 mg/l and did not show any significant variations with depth (Figure 7). Although not directly observed, sulphate reduction is assumed to have occurred at non-detectable levels during Stage 2, as it is known to take place in parallel or even before iron reduction, dependent on the local geochemical setting (Postma and Jakobsen 1996).

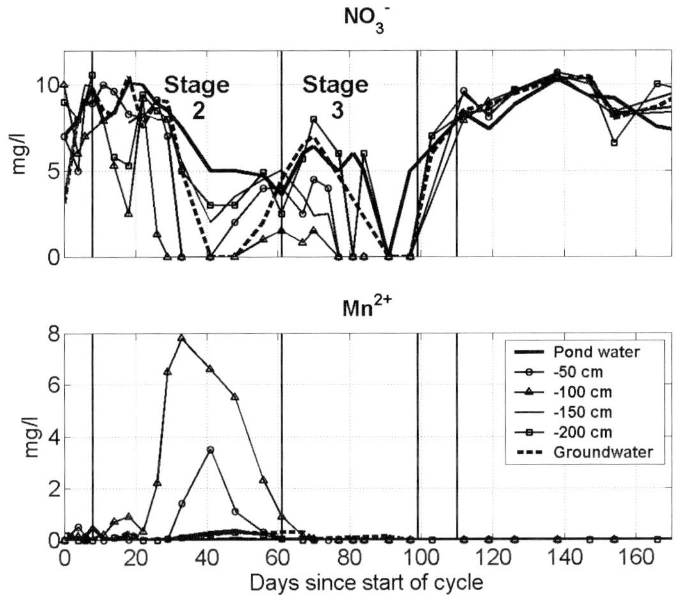


**Fig. 6.** Nitrate (as $NO_3^-$) and manganese ($Mn^{2+}$) in the pond and groundwater and at depths of 50 cm, 100 cm, 150 cm and 200 cm beneath the pond.

During the rapid change from water saturated (Stage 2) to water unsaturated conditions (Stage 3), oxygen concentrations increased up to 5 mg/l (Figure 5). Although aerobic respiration seems to be the dominant redox reaction during Stage 3, nitrate concentrations at 3 suction cups were still somewhat lower than in the surface water (Figure 6). These patchily distributed zones of more reducing conditions (also observed in Stage 2) are considered to result from (i) the formation of anaerobic microsites caused by particulate organic carbon contained in the sediment (e.g., Murphy and Schramke 1998, Hamersley and Howes 2002) and (ii) non-uniform flow caused by the physical heterogeneity of the sediment, i.e., hydraulic conductivity. Concomitant with the appearance of atmospheric oxygen (during the transition from Stage 2 to Stage 3) extremely high sulphate concentrations of up to 370 mg/l were observed in the water extracted from three suction cups. However, in the following 20 days they declined rapidly and returned to background concentrations (Figure 7). The observed peaks of sulphate concentrations are most likely caused by the rapid oxidation of iron sulphide minerals that had formed previously (during Stage 2) as a result of sulphate reduction. However, rapid oxidation of sulphide minerals due to water table fluctuations has been observed for other, comparable hydrogeochemical systems (e.g., Sinke et al. 1998). Concomitant with the peaks of sulphate concentration, very high concentrations of calcium were observed at the same locations below the pond (data not shown). As sulphide oxidation produces protons, the heightened calcium concentrations apparently resulted from pH buffering by the dissolution of calcite.

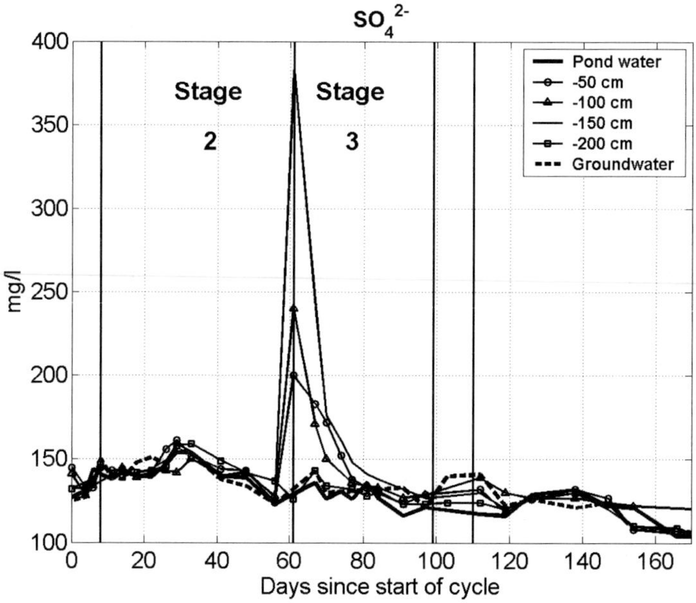


**Fig. 7.** Sulfate ($SO_4^{2-}$) concentrations in the pond and groundwater and at depths of 50 cm, 100 cm, 150 cm and 200 cm beneath the pond.

## Conclusions

The field study reported here shows that the highly dynamic hydraulic system beneath the recharge pond has a considerable impact on the spatial and temporal distribution of different redox conditions. In particular, three major conclusions can be drawn. First of all, the entire hydraulic system is controlled by the formation of a clogging layer, which results in alternating saturated/unsaturated conditions below the pond. Secondly, the sudden change from water saturated to water unsaturated condition produced remarkable changes of the local hydrogeochemistry. Finally, the existence of anaerobic microsites resulting from chemical and/or physical heterogeneity probably caused the non-uniform redox pattern observed below the pond.

## Acknowledgements

The authors would like to thank Berliner Wasserbetriebe and Veolia Water for financial support of the NASRI project at the Kompetenzzentrum Wasser Berlin making this study possible. Henning Prommer was supported by CORONA, a 5$^{th}$ framework Project of the European Union. Special thanks also to Grit Siegert, Kilian Pramschiefer, Elke Weiß, Silke Meier, Doreen Richter and Achim Bartels for their technical help, and Peter Franzmann for linguistic improvements of the manuscript.

## References

Agertved J, Rügge K, Barker JF (1992) Transformation of the herbicides MCPP and atrazine under natural aquifer conditions. Ground Water 30(4):500-506

Asano T (1992) Artificial Recharge of Groundwater with Reclaimed Municipal Wastewater: Current Status and Proposed Criteria. Water Science and Technology 25(12):87-92

Bosma TNP, Ballemans EMW, Hoekstra NK, te Welscher RAG, Smeenk JGMM, Schraa G, Zehnder AJB (1996) Biotransformation of Organics in Soil Columns and an Infiltration Area. Ground Water 34(1):49-56

Bouwer H (2002) Artificial recharge of groundwater: hydrogeology and engineering. Hydrogeology Journal 10(1):121-142

Bouwer H (1991) Role of groundwater recharge in treatment and storage of wastewater for reuse. Water Science and Technology 24(9):295-302

Bouwer H (1978) Groundwater Hydrology. McGraw-Hill Book Company, New York

Brun A, Broholm K (2001) Basin infiltration at Arrenæs site, Denmark: Tracer test and quality changes during aquifer passage below infiltration basins. In: Artificial recharge of groundwater, Final Report, European Commission Project ENV4-CT95-0071, pp 177–181

Champ DR, Gulens J, Jackson RE (1979) Oxidation-reduction sequences in groundwater flow systems. Can J Earth Sci 16:12-23

Eckert P, Appelo CAJ (2002) Hydrogeochemical modeling of enhanced benzene, toluene, ethylbenzene, xylene (BTEX) remediation with nitrate, Water Resour Res 38(8):10.1029/2001WR000692

Fujita Y, Zhou J, Orwin E, Rheinhard M, Davisson ML, Hudson GB (1998) Tracking the movement of recharge water after infiltration. In: Peters JH et al. (ed) Artificial recharge of groundwater, AA Balkema, Amsterdam, pp 155-160

Greskowiak J, Prommer H, Massmann G, Johnston CD, Nützmann G, Pekdeger A (2005) The Impact of Variably Saturated Conditions on Hydrogeochemical Changes during Artificial Recharge of Groundwater, Applied Geochemistry, in press

Grützmacher G, Böttcher G, Chorus I, Knappe A, Pegdeker A (2002) Cyanobacterial toxins in bank filtered water from Lake Wannsee, Berlin. In: Dillon PJ (ed) Management of Aquifer Recharge for Sustainability, AA Balkema, Lisse, pp 175-179

Hamersley MR, Howes BL (2002) Control of denitrification in a septage-treating artificial wetland: the dual role of particulate organic carbon. Wat Res 36:4415-4427

Hecht H, Kölling M (2001) A low-cost optode array measuring system based on 1mm plastic optical fibers – new technique for in situ detection and quantification of pyrite weathering processes. Sensors and Actuators B 81:76-82

Herczeg AL, Rattray KJ, Dillon PJ, Pavelic P, Barry KE (2004) Geochemical Processes During Five Years of Aquifer Storage Recovery. Ground Water 42(3):438-445

Holm JV, Rügge K, Bjerg PL, Christensen TH (1995) Occurrence and Distribution of Pharmaceutical Organic Compounds in the Groundwater Downgradient of a Landfill (Grindsted, Denmark). Environ Sci Technol 29(5):1415-1420

Johnson JS, Baker LA, Fox P (1999) Geochemical transformations during artificial groundwater charge: Soil-water interactions of inorganic constituents. Water Res 33(1):196-206

Långmark J, Storey MV, Ashbolt NJ, Stenstöm TA (2004) Artificial groundwater treatment: biofilm activity and organic carbon removal performance. Water Res 38(1):740-748

Lindroos AJ, Kitunen V, Derome J, Helmisaari HS (2002) Changes in dissolved organic carbon during artificial recharge of groundwater in a forested esker in Southern Finland. Water Res 36(20):4951-4958

Murphy EM, Schramke JA (1998) Estimation of microbial respiration rates in groundwater by chemical modeling constrained with stable isotopes. Geochim Cosmochim Acta 62(21/22):3395-3406

Pekdeger A, Massmann G, Ohm B, Pühringer S, Richter D, Engemann N, Gruß S (2002) Hydrogeological-hydrogeochemical processes during bank filtration and ground water recharge using a multi tracer approach. 1st NASRI report, KompetenzZentrum Wasser Berlin, pp 14-50

Postma D, Jakobsen R (1996) Redox zonation: equilibrium constraints on the Fe (III)/SO4-reduction interface. Geochim Cosmochim Acta 60:3169-3175

Prommer H, Stuyfzand PJ (2005) Identification of temperature-dependent water quality changes during a deep well injection experiment in a pyritic aquifer. Accepted for publication in Environ Sci and Technol

Quanrud DM, Arnold RG, Wilson LG, Gordon HJ, Graham DW, Amy GL (1996) Fate of organics during column studies of soil aquifer treatment. J Environmental Engineering 22(4):314-321

Ray C, Melin G, Linsky RB (2002) Riverbank filtration, Water Science and Techology Library, vol 43. Kluwer Academic Publishers, Netherlands

Sinke AJC, Dury O, Zobrist J (1998) Effects of a fluctuating water table: column study on redox dynamics and fate of some organic pollutants. J Cont Hydrol 33:231-246

Stuyfzand PJ. (1998) Quality changes upon injection into anoxic aquifers in the Netherlands: Evaluation of 11 experiments. In: Peters JH (ed) Artificial recharge of groundwater, AA Balkema, Amsterdam, pp 283-291

# Transport of Cr(VI), Ni(II) and Mn(II) through metallurgical wastes. Batch and column experiments

R. Rodríguez, L. Candela

Department of Geotechnical Engineering & Geosciences. School of Civil Engineering - UPC -. Gran Capitán s/n, Edif. D-2. Barcelona 08034, Spain.

Corresponding author: Jordi Girona 1-3, Modulo D-2. Campus Nord 08028. Barcelona. Spain. Phone number: +34.93.4016868, Fax number: +34.93.4017251, e-mail: lucila.candela@upc.edu

## Abstract

The adsorption/ desorption processes in metallurgical wastes of Cr, Ni and Mn in single, binary and ternary solutions were investigated using batch tests and miscible displacement column experiments. For the experiments, mine wastes from Moa (Cuba) composed by oxi(hydroxi) iron minerals and heavy metals deposited in tailing dams were used. The adsorption isotherms of the three metals are strongly non-linear and are well described by the Freundlich isotherm. Desorption isotherms are linear. The adsorption/desorption process of Cr, Ni and Mn shows hysteresis. The adsorption of metals can be attributed to the favourable electrostatic interaction due to mineralogical composition of the wastes. The shape of breakthrough curves from column experiments provide evidence that transport of Ni, Mn and Cr is non-ideal with asymmetry and great tailing, showing an important retardation relative to a tracer. When using a ternary solution (Cr, Ni, and Mn), retardation of Ni and Mn was reduced in comparison with the single solute experiment, consequence of the Ni and Mn competition for adsorption sites, while Cr did not show any change. In the Ni-Mn binary solution column transport experiment, Mn adsorption was more rapidly attained than Ni; and both metals show different adsorption behaviour, when adsorption of Ni increases, sorption of Mn decreases. One more aspect being investigated was the sorption capacity of the metallurgical waste. According to the mass-balance analysis of the results, the waste shows a very high effective sorption.

## Keywords

Flow and transport of metals; metallurgical waste; sorption and desorption; sorption hysteresis

## Introduction

The fate of heavy metals emanating from mining, metallurgical, chemical industry and hazardous waste sites (tailing dams, landfills) has become a major water pollution issue. Among these activities mining, metallurgical and chemical industry produce a great volume of aqueous effluents. Common metals found in waste water are Cr, Fe, Mn, Co, Ni, Cu, Zn, Cd, Hg and Pb which need to be treated in order to avoid water pollution. The mobility of trace metals in soils and sediments is of particular concern because of potential bioaccumulation. Although metals may be strongly sorbed to soils, clays and oxides, does not guarantee immobilisation of contaminant metals in the subsurface. Metal transport in the subsurface environment is strongly influenced by chemical processes, such as complexation, sorption and dissolution. As known from literature, heavy metals can be adsorbed by soil colloids and clay minerals such as amorphous iron oxides, allumina, kaolinite and montmorillonite (Griffin et al. 1964). The adsorption-desorption process in both natural soils and different minerals plays a significant role controlling the aqueous concentration of heavy metal species in soils (Rait et al. 1989, Buchter et al. 1989, Adriano 1995, Selim and Amacher 1997, Khaodhiar et al. 2000). However, little is known about the mode of occurrence and behaviour of heavy metals in adsorption processes by metallurgical wastes (Apak et al. 1998, Rodríguez 2002). The main objective of this work is to present the adsorption-desorption behaviour of Cr, Ni and Mn in (oxi)hydroxy metallurgical wastes from tailing dams.

## Materials and Methods

Tailing dam wastes from the nickel-cobalt metallurgical processes of Moa (Cuba) were selected because of being representative of mining activities in the overall area (Rodriguez 2002). Ten waste samples of one kilo each were collected at 10-20 cm depth during a sampling campaign in April 1996. The samples were transferred to polyethylene bags for transportation and dried at 60°C, prior to any experimentation in the laboratory.

Batch equilibrium sorption isotherms for Cr(VI), Ni(II) and Mn(II) were determined at 22±2°C after an equilibration period of 24 hours. All experiments were conducted in a 0.01 mol·$L^{-1}$ of $KNO_3$ electrolyte solution. The standard solutions of metals used in the experiment were prepared using stock standard solutions of $K_2CrO_4$, $Ni(NO_3)_2$ $6H_2O$ and $Mn(NO_3)_2$. The liquid-solid ratio in all experiments was 1:10.

For the charge equilibration of the solid particles, solid waste samples of 2 g were prepared in pre-weighted polypropylene tubes, after the addition of 20 ml of 1.0 mol·$L^{-1}$ $KNO_3$ they were shaken between 5 and 1440 minutes. Suspensions were centrifuged at 900 rpm, and the supernatants discarded. After equilibration, and for the adsorption experiments, the tubes containing waste and residual solution were weighted for residual solution calculation. Aliquots of 20 ml, containing different initial concentrations of metals, prepared with 0.01 mol·$L^{-1}$ $KNO_3$, were then added to the tubes, and shaken for 5, 10, 30, 60, 120 160, 480 and 1440 minutes. The suspensions were centrifuged at 900 rpm for 5 minutes, and the supernatants were decanted and collected for pH and metals content determination. For the desorption experiments, aliquots of 20 ml of electrolytic solution, prepared with 0.01 mol·$L^{-1}$ $KNO_3$, were then added to the tubes containing the wastes, and shaken (5 to 4320 minutes). The suspensions were centrifuged at 900 rpm for five minutes, and the supernatants were decanted and collected for pH and metal analysis.

To adjust the solution to the desired pH acid or basic solution (0.1 mol.$L^{-1}$ HCl or NaOH) was added. After 24 hr of equilibration time, suspensions were centrifuged and pH determined.

In all tests the effluent was filtered through 0.45 μm filters and metal concentration in the effluent was determined by ICP-MS. The adsorbed (Sa) and desorbed (Sd) metal and effective adsorption (Se) of wastes was estimated according to the following equations,

$$Sa = \frac{(C_o - C_w)V}{M} \tag{1}$$

$$Se = \frac{(C_{wf} - C_{wi})V}{M} \quad (2)$$

$$Sd = Sa - Se \quad (3)$$

where $C_o$ is the input concentration in the initial solution, $C_w$ is aqueous concentration, $C_{wi}$ is aqueous concentration in initial solution used for desorption process, $C_{wf}$ is aqueous concentration in the final solution after desorption, M is the total mass of waste and V is the electrolytic solution volume.

For the miscible displacement experiments, wastes were packed into PVC columns (Table 1). One dual HPLC pump was connected to the columns and 6 port plastic valves were used to switch between the different solutions of interest. Air dried waste samples were packed in the column under vibration and progressively saturated with a diluted electrolytic solution of 0.01 mol·$L^{-1}$ $KNO_3$ until saturated steady-state flow regime was established. The effluent flow rate was periodically monitored in order to ensure steady state flow. Pentafluorbenzoate-PFB was the tracer used in the column test to verify physical characteristics of the column (concentration was 5 mg/L and volume input 60 mL). The procedure with single metal solution included the initial saturation of the column in order to obtain maximum adsorption capacity of the waste before to start the desorption exercise.

**Table 1.** Miscible displacement experiments. Waste column characteristics

| Column characteristics | | | |
|---|---|---|---|
| Length (L) cm | 10.00 | Pore volume (Vp) $cm^3$ | 12.22 |
| Diameter (Φ) cm | 1.60 | Porosity (η) $cm^3/cm^3$ | 0.61 |
| Total volume (V) ) $cm^3$ | 10.55 | Volumetric water content (θ) $cm^3/cm^3$ | 0.61 |
| Weight of waste (M) g | 15.66 | Flow velocity (v) cm/h | 1.20 |
| Bulk density (ρh) $g/cm^3$ | 2.17 | Dead volume (Vm) $cm^3$ | 0.12 |
| Particle density (ρs) $g/cm^3$ | 3.97 | | |

# Results

## Metallurgical Solid Wastes

Results from laser diffraction analysis show that wastes are mainly composed by silt-size particles, with some small clay-size particles. Grain size distribution is poorly graded and average diameter is about 0.02 mm. According to XR, the main minerals present are Fe and Al (oxi)hydroxides, primary magnetite and hematite, being the Fe content greater that 40% of the total weight. Chromite and chrome spinel is the major form of Cr appearance. Table 2, shows some physico-chemical characteristics of the metallurgical wastes.

**Table 2.** Principal characteristics of the metallurgical wastes (average from 5 samples)

| Characteristics | | | |
|---|---|---|---|
| Sand particles >2 mm (%) | 10 | CEC* (meq/100g of solid) | 10 |
| Silt particles 2 mm - 0.002 mm (%) | 70 | Organic matter (OM) (%) | 4.6 |
| Clay particles < 0.002 mm (%) | 20 | Amorphous Fe(g/kg) | 21.321 |
| Specific surface ($m^2/g$) | 80-112 | Amorphous Mn (g/kg) | 2.33 |
| pH (ratio 1:2.5) | 6.45 | Amorphous Al (g/kg) | 3.015 |
| Electric conductivity (mS/cm) (ratio 1:2.5) | 670 | Amorphous Si (g/kg) | 0.208 |

*Cation Exchange Capacity

## Batch Experiments

### *Sorption. Single solution*

To analyse heavy metals removal and time dependence, a sorption test was initially performed for different time periods ranging from 5 to 1440 minutes. Results of batch experiments of adsorption kinetics indicated that a period of 8 hours was enough to attain equilibrium for the three metals and, 24 hours were chosen as the equilibrium time for all batch experiments (Figure 1a). Equilibrium in desorption process apparently occur in less than 5 minutes (Figure 1b).

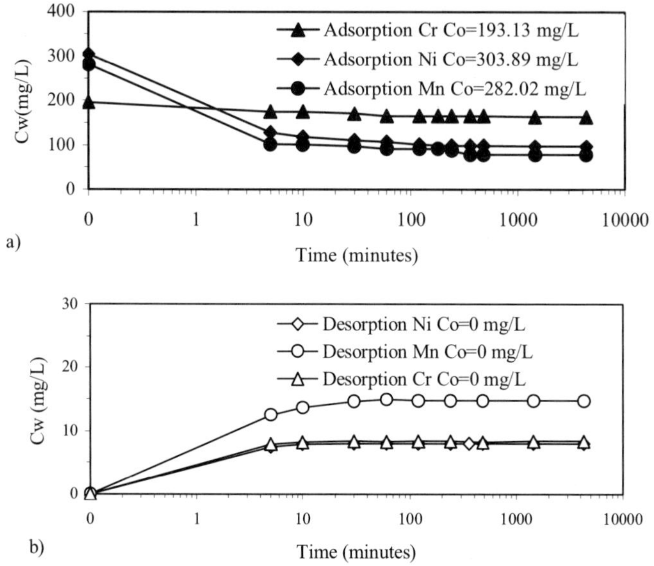


**Fig. 1.** The adsorption/desorption kinetics of Ni, Mn and Cr

The effect of pH in the sorption process of metals is shown in Figure 2. Nickel and Manganese adsorption increases with pH, while chromium adsorption is greater at a low pH and it decreases with increasing pH values. This effect is mainly controlled by an electrostatic adsorption process.

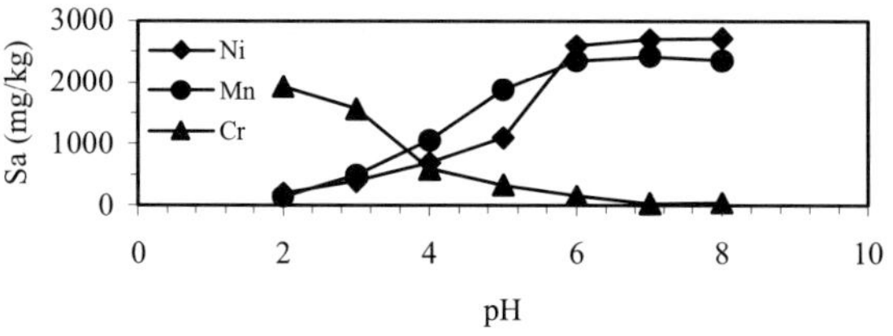


**Fig. 2.** Ni, Mn and Cr adsorption dependence on pH

Adsorption-desorption isotherms for the three metals are shown in Figure 3. All curves show a gradual decrease of waste adsorption with regard to the aqueous concentration as a consequence of active surface sites filled with metal. Isotherms for Ni and Mn show an H-type shape, although differences in the amount of metal absorbed are observed; Cr(VI) was less strongly adsorbed by the waste. The H-type isotherm is indicative of strong adsorbate-adsorptive interactions such as inner-sphere complexes (Giles et al. 1960, Sparks 1995). Total adsorption of metals in the waste is only produced when solution concentration is lower than 0.2 mM (Cr), 2.8 mM (Ni) and 2 mM (Mn) respectively.

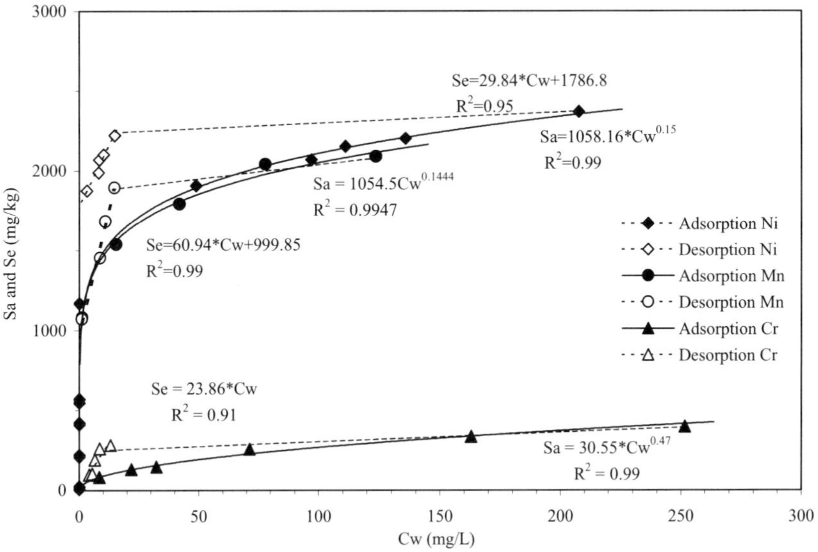


**Fig. 3.** Adsorption-desorption isotherms for Ni, Mn and Cr. Sa: mg/kg of adsorbed metal; Se: mg/kg of desorbed metal

Sorption isotherms for the three metals are non-linear and are well described by the Freundlich equation (4). Presence of non linear adsorption isotherm implies that retardation maybe a function of metal concentrations in the solution.

$$S = K_d C_w^n \tag{4}$$

As seen in Figure 3, the Freundlich equation was fitted to the experimental adsorption data giving satisfactory correlation coefficients (r>0.91).

The desorption isotherms of the three metals are linear and well described by the following equation,

$$S = K_d C_w . \tag{5}$$

According to Figure 3 sorption was not completely reversible, as some Ni, Mn and Cr(VI) was desorbed. The $K_d$ value in all adsorption experiments was usually of one order of magnitude higher than desorption, showing the irreversibility of metal adsorption. The observed sorption hysteresis is typical for trace metals adsorption-desorption phenomena in soils (e.g. Weng et al. 1994, Spark 1995). The difference between the absorbed amount of Ni and Mn with regard to Cr(VI) is controlled by the pH.

Results from literature show that the $K_d$ values for Ni, Mn and Cr of the Moa waste is much greater than those observed in common natural soils, minerals and synthetic materials (Selim and Amacher 1997, Buchter et al. 1989, Rodríguez 2002, Rodríguez et al. 2004).

## *Adsorption Isotherms of Cr, Ni and Mn in a ternary solution*

In order to compare the single solute experiment behaviour with a more real metal transport, a ternary metal solution of Cr, Ni, and Mn was used for the kinetic experiments. In the ternary solution, total adsorption of Ni and Mn is much lower as a consequence of the Ni and Mn competition for adsorption sites, while Cr did not show any significant change.

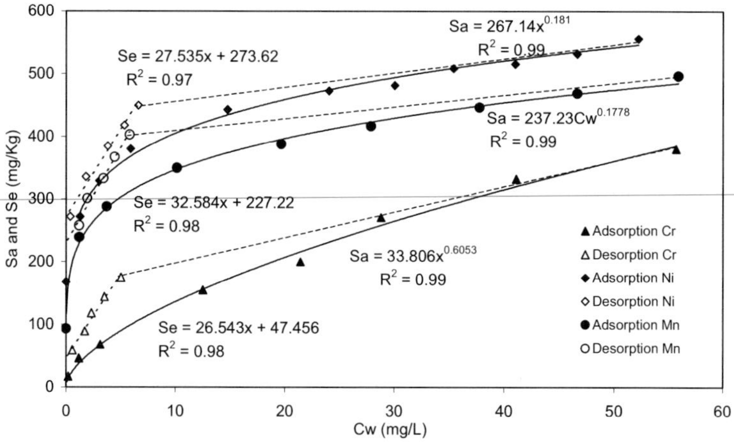


**Fig. 4.** Adsorption isotherm of Ni, Mn and Cr in the waste from a ternary metal solution. Sa: mg/kg of adsorbed metal; Sd: mg/kg of desorbed metal

**Table 3.** Adsorbed mass in batch tests. Sa: maximum adsorbed mass; Se: effective adsorbed mass; Sd: desorbed mass

| | Single solution | | | | Ternary solution | | | |
|---|---|---|---|---|---|---|---|---|
| | Adsorption-Desorption | | | | Adsorption-Desorption | | | |
| Metals | Cw (mg/L) | Sa (mg/kg) | Se (mg/kg) | Sd (mg/kg) | Cw (mg/L) | Sa (mg/kg) | Se (mg/kg | Sd (mg/kg) |
| Mn(II) | 332.53 | 2100 | 999 | 1101 | 162 | 556 | 227 | 329 |
| Cr(VI) | 291.43 | 400 | 100 | 300 | 167 | 370 | 47 | 323 |
| Ni(II) | 379.77 | 2300 | 1786 | 1514 | 162 | 556 | 273 | 283 |

The three metals present a non-linear isotherm and are well described by the Freundlich equations, being $K_d$ for Ni and Mn about four fold of the value obtained in the single solution. Desorption is linear, and for the three metals the sorption process presents important hysteretic behaviour (Figure 4). Also mass adsorption is important.

## Adsorption-desorption miscible displacement experiments

### *Tracer and three single-metal solution tests*

Measured breakthrough curves (BTCs) for the three single-metal experiments (Ni, Mn and Cr) in the waste are reported in Figure 5. The BTCs were asymmetrical and showed significant tailing, with respect to the tracer test. The most dominant effect detected in the BTCs is the important effective mass absortion in the waste as seen in Table 4.

The results of mass balance analysis of the BTCs showed that approximately 98.4% of Cr(VI), 96% of Ni and 96% of Mn, entering the column was recovered at the end of the flow experiment. It is important to notice that the concentration of metal in the effluent did not reach zero at the end of the flow experiments, but final concentration was under the detection limit. This fact can be attributed to two factors, one being the use of a monovalent electrolytic solution (0.01mol.L$^{-1}$ $KNO_3$) and also to the chemisorption of metals. The observed hysteretics is in good agreement with the results from batch tests.

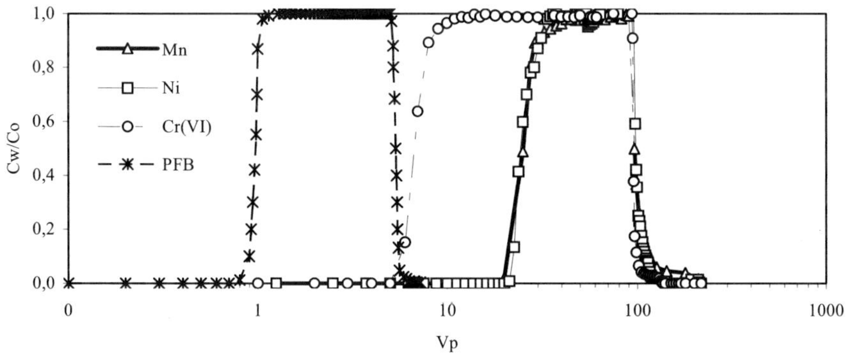


**Fig. 5.** Single metal solution (Ni, Mn and Cr) and PFB tracer breakthrough curves.

### Test with a ternary metal solution

When performing the experiment with a ternary solution of Cr, Ni, and Mn, retardation of Ni and Mn is greatly reduced due to the existing competition for adsorption sites as also happened in the batch tests, while Cr did not show an important change. The Ni and Mn desorption process was linear.

The existence of a very high effective sorption (502 mg/kg of Mn, 1020 mg/kg of Ni and 385 mg/kg of Cr) is still detected when the experiment is carried out with the three metals in solution. Differences in mass sorption for the single and ternary experiments are shown in Table 3.

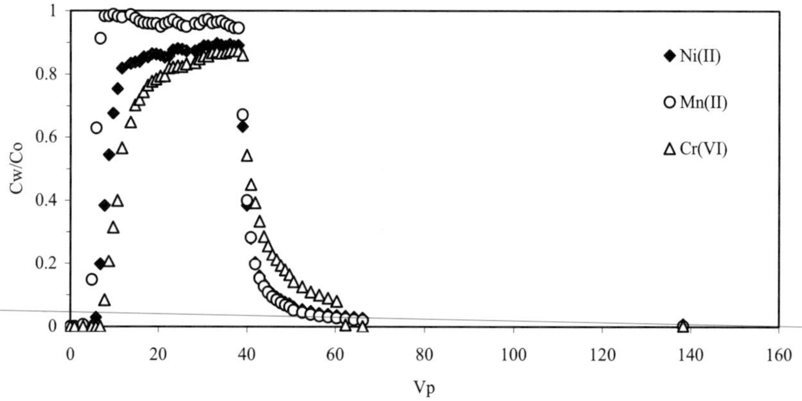


**Fig. 6.** Breakthrough curves for the ternary metal solution (Ni, Mn and Cr)

### Test with a binary metal solution

When the displacement experiment is performed with a single or ternary solution, breakthrough curves of Ni and Mn overlap due to their similar sorption properties (Figures 5 and 6). In order to asses Mn and Ni adsorption behaviour, a flow and transport test with a binary Ni-Mn solution was performed. The BTCs are asymmetrical with significant tailing in both metals and adsorption of Mn was more rapidly attained than Ni, which also happened in the ternary solution. However, when reaching 10 pore volume (Vp) approximately, adsorption of Ni increases while the tendency of Mn is to decrease (Figure 7). The effective adsorbed mass obtained is the same as the obtained for the ternary experiment.

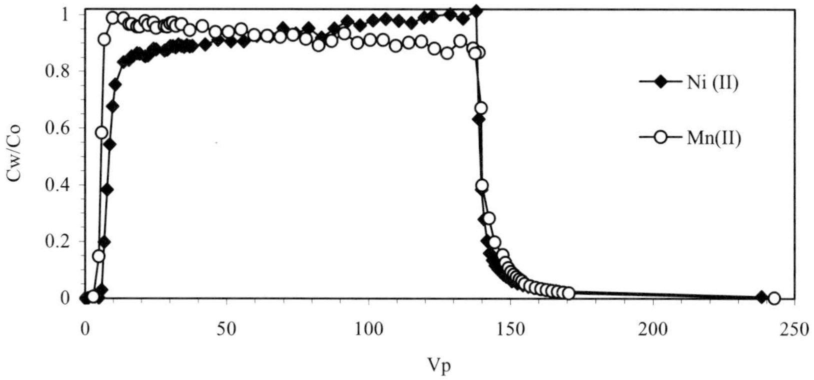


**Fig. 7.** Breakthrough curves for a binary metal solution of Ni and Mn

**Table 4.** Adsorbed mass. Maximum mass adsorbed (Samax) and effective adsorbed mass (Se)

| Metal | Single metal solution | | | Ternary metal solution | | | Binary metal solution | | |
|---|---|---|---|---|---|---|---|---|---|
| | Co˙ | Samax | Se | Co | Smax | Se | Co | Smax | Se |
| | mg/L | mg/kg | | mg/L | mg/kg | | mg/L | mg/kg | |
| Mn(II) | 267 | 2285 | 1981 | 419.8 | 864 | 502 | 419.78 | 864 | 502 |
| Cr(VI) | 277 | 624 | 295 | 103.5 | 560 | 385 | -- | -- | -- |
| Ni(II) | 416 | 2744 | 2168 | 336.84 | 1381 | 1020 | 336.84 | 1381 | 1020 |

## Discussion

The shape of breakthrough curves from column experiment with single, binary and ternary metal solution provide evidence that transport of Cr, Ni and Mn is non linear being retardation dependent of metal concentration in solution. The adsorption capacity of Ni and Mn is significantly affected by the presence of a metal with similar sorptive properties which entails competition for the adsorption sites.

Sorption is not completely reversible and the high sorption is dependent on the metal oxide and hydroxides content. As the waste shows an important presence of iron oxides and ferrihydrite mineral particles, heavy metals adsorption is enhanced by the important specific surface and the electrostatic effects originated by the negative charge under the existing alkaline pH conditions. The adsorbed amount of Ni, Mn and Cr(VI) is controlled by the pH values of the medium and the electrostatic charge of the mineral particles present in the waste. Chromium adsorption shows greater adsorption at a low pH and it generally decreases with increasing pH values of the system. The Cr(VI) adsorption results agree with the results reported for iron-rich soils (Zachara et al. 1988, Weng et al. 1994, Khaodhiar et al. 2000).

## Conclusions

Data from batch studies proved that adsorption of Ni and Mn by the waste was up to 10 fold greater than Cr adsorption and also suggested a limited mobility of Ni, Mn and Cr through the metallurgical waste. The initial Ni, Mn and Cr concentration in the solution has been found to be an important parameter for the removal of metals from solution. The $K_d$ value of Ni, Mn and Cr in the waste is more than one order of magnitude than in common natural soils, minerals and synthetic materials. Considering the high capacity of adsorption of the metallurgical waste used in this experiment, more experiments need to be done to asses the possibility of future use as a sorbent material for Ni, Mn and Cr(VI) removal from contaminated water.

## Acknowledgments

This work was done under the framework of the Spanish National Project (CICYT) PPQ2001-2100-C04 and PB/44/FS/2002 financed by Fundación Séneca, Murcia, Spain.

## References

Adriano DC (1995) Trace elements in the terrestrial environment. Springer-Verlag New York

Apak R, Tütem E, Hügül M, Hizal J (1998) Heavy metals cation retention by un conventional sorbents (red muds and fly ashes). Water Research 32:430-440

Buchter B, Davidoff B, Amacher MC, Hinz C, Iskandar IK and Selim HM (1989) Correlation of Freundlich $K_d$ and n retention parameters with soils and elements. Soil Science 148:370-379

Giles C, MacEwan TH, Nakhwa SN, Smith D (1960) A system of classification of solution absorption isotherms, and its use in diagnosis of adsorption mechanisms and in measurement of specific surface areas of solids. J Chem Soc 3:3973-3993

Griffin RA, Au AK, Frost RR (1964) Effect of pH on adsorption of chromium from landfill leachate by clay minerals. J Environ Sci Health, Part A, A12:431-449

Khaodhiar S, Azizian A, Osathaphan Y, Nelson PO (2000) Copper, chromium and arsenic adsorption and equilibrium modelling in an iron-oxide-coated sand, background electrolyte system. Water, Air and Soil Pollution 119:105-200

Rait D, Eary LE, Zachara JM (1989) Environmental chemistry of chromium. Sci Tot Environ 86:15-23

Rodríguez R (2002) Estudio Experimental de flujo y transporte de cromo, níquel y manganeso en residuos de la zona minera de Moa (Cuba): influencia del comportamiento hidromecánico. Ph.D. Thesis, Technical University of Catalonia (UPC), Barcelona, Spain. 460. http://www.tdcat.cesca.es/TDCat-0731102-084652/

Rodríguez R, Hidalgo M, Salvadó V, Candela C, Lloret A, Queralt I (2004) The use of metallurgical wastes in reactive barriers. Proceeding International RILEM conference on the use of recycled materials in buildings and structures. Volume 2, RILEM Publications s.a.r.l. Bagneux-France, pp 1115-1125

Selim HM, Amacher C (1997) Reactivity and transport of heavy metals in soils. Lewis Publishers, New York

Sparks DL (1995) Ion exchange processes, Chapter 6 in Environmental soil chemistry. Academic press, New York, pp 141-158

Weng CH, Huang CP, Allen HE, Cheng AHD, Sanders PF (1994) Chromium leaching behaviour in soil derived from chromite ore processing waste. The Science of the Total Environment 154:71-86

Zachara JM, Cowan CE, Schmidt, Ainsworth RL (1988) Chromate adsorption on kaolinite Clay Clay Minerals. 36:317-326

# Modeling adsorption-desorption processes of Cu on montmorillonite and the effect of competitive adsorption with a cationic pesticide

T. Undabeytia[1*], S. Nir[2], G. Rytwo[3], C. Serban[2], E. Morillo[1], C. Maqueda[1]

[1)] Institute of Natural Resources and Agrobiology, IRNAS (CSIC). Apdo 1052. 41080 Seville, Spain
[2)] Faculty of Agriculture, HUJI, Rehovot 76100, Israel
[3)] Tel Hai Academic College, Upper Galilee 12210, Israel

*Corresponding author: Avda. Reina Mercedes 10. Tel: +34-954624711, fax: +34-954624002, e-mail: undabeyt@irnase.csic.es

## Abstract

The effect of the ionic strength on adsorption of Cu on Ca-montmorillonite (SAz-1) was studied at concentrations ranging from 31 to 516 μM. An adsorption model was employed in the analysis of the data. When the background electrolyte was $NaClO_4$ the ionic interchange was suppressed at 0.5 M, and Cu adsorption was taking place on edge sites, reaching a plateau at about 24 mmol/kg. A further increase in ionic strength did not make any effect on Cu adsorption, suggesting that the heavy metal was being adsorbed by inner sphere complexes on the edge sites. When the electrolyte used was NaCl the amounts of Cu adsorbed were reduced. The model predicted well the adsorption data by considering adsorption of $CuCl^+$ species. Adsorption-desorption processes of Cu on Ca-montmorillonite in media of 0.01 and 0.1 M NaCl showed hysteresis. Model calculations predict the desorbed amounts fairly well. According to the model the hysteresis is mainly attributed to the heterogeneity of sites for the adsorption of Cu. The hysteresis arising from the planar sites is largely due to reduced competition for adsorption and enhancement in the magnitude of the surface potential. The presence of the cationic pesticide chlordimeform reduces strongly the sorption of the metal on the planar positions unlike the edge sites.

However, Cu sorption increases on the clay treated previously with chlordimeform which was due to the opening of the clay platelets after some molecules of the pesticide are adsorbed, facilitating the subsequent penetration of the metal and its adsorption on planar positions. This cooperative adsorption was due to the fact that the loading of the pesticide on the clay was a very small amount of the CEC.

## Keywords

Montmorillonite; copper; chlordimeform; modeling; adsorption; desorption; hysteresis.

## Introduction

Copper content in soils has lately increased as a result of repetitive application of manure, Cu-containing pesticides, sludges, and industrial mine wastes. In this context, adsorption-desorption of copper on laminar silicates is of prime importance since their surfaces show a high affinity for fixation of heavy metals (Zachara et al. 1992).

Studies on the adsorption of the heavy metals on layer silicates have proposed the existence of sites of different affinity, where the higher affinity sites occupy a small fraction of the surface (Undabeytia et al. 1996). The increase in the number of these high affinity sites with increasing pH suggests that they are associated with the edge sites of the silicate layer whereas those of lower affinity are ionic interlayer sites (Zachara and McKinley 1993). However, the contribution of these sites to copper adsorption-desorption on montmorillonite has not yet been well established.

In the current work, a model of cation adsorption in closed system (Nir 1986) is further extended to consider adsorption/desorption of Cu (or other cations) on two types of sites of a Ca-montmorillonite. The employment of the model and experimental tests permits us to critically evaluate and explain the hysteresis phenomenon. Another aspect emphasized in this study is the influence of the presence of the pesticide chlordimeform on Cu adsorption since in general, little information is available on processes taking place when heavy metals and pesticides are present together.

## Materials and Methods

### Adsorption-desorption procedures

The clay mineral used was a standard montmorillonite from Arizona (SAz-1). The adsorption experiments were done in triplicate by mixing 0.1 g of clay mineral with 20 ml of solutions containing various Cu and/or chlordimeform concentrations. The range of concentrations used was from 31 to 516 μM for Cu and from 100 to 1000 μM for chlordimeform. The experiments were carried out in $NaClO_4$ or NaCl media, the ionic strengths used being 0.01, 0.1, 0.5 and 1.0 M. The samples were shaken for 24 h at 20 ± 1 °C. Desorption experiments were performed after adsorption equilibrium was reached, by removing half of the supernatant after centrifugation, replacing it with 10 ml of the initial medium and ionic strength, allowing equilibration for an additional 24 h period, and subsequently continuing as in the adsorption experiment. This process was repeated twice more. In some desorption points, the removed supernatant was replenished with the solution of the background electrolyte including the amounts of Ca and Mg present in the volume removed.

Cu adsorption on SAz-1 was determined at the equilibrium pH 3.5 by adjusting the initial solutions with aliquots of 0.1 M HCl. Adsorption of the metal was also carried out on montmorillonite treated previously with chlordimeform solutions as indicated in the adsorption procedure.
Chlordimeform adsorption was also performed on a standard Na-montmorillonite (SWy-1).

### Model calculations

The cation adsorption model used in this study has been described in Nir (1986). The program considers cation binding and the electrostatic Gouy-Chapman equations, and solves iteratively for the solution concentrations of all cations in a closed system. Cation adsorption is obtained by considering specific binding to surface sites and residence in the double layer region. The model was extended to account explicitly for complexation reactions in solution. In the current study we extended the model to account for cation adsorption on two types of sites, planar (P) and edge (E). For brevity we only present a few equations and define the model parameters.

The symbols $M_i^+$ and $M_j^{2+}$ denote cations that can bind to singly charged negative sites, $P^-$ or $E^-$, on the planar and edge surfaces of the clay mineral, respectively. The binding reaction for a monovalent cation is:

$$P^- + M_i^+ \Rightarrow (PM_i)^0 \quad (1)$$

$$E^- + M_i^+ \Rightarrow (EM_i)^0 \quad (1')$$

The respective binding coefficients will be denoted by Ki and $Ki_E$, and are given by:

$$Ki = [(PM_i)^0]/([P^-][M_{i(0)}^+]) \quad (2)$$

in which $[Mi_{(0)}^+]$ is the concentration of the cation at the surface. Similarly, $Ki_E$ is given by an expression which includes $(EM_i)^0$ and $E^-$.

Divalent cations can react with one surface site yielding a charged complex, or they can associate to form a neutral complex, by binding to a divalent charged site $(P^{2-})=(P^-)/2$. For neutral complexes:

$$P^{2-} + M_j^{2+} \Rightarrow (P^{2-}M_j^{2+})^0 \quad (3)$$

and similarly,

$$E^{2-} + M_j^{2+} \Rightarrow (E^{2-}M_j^{2+})^0 \quad (3')$$

with $K_2j$ and $K_{2E}j$ defined similarly to Ki and $Ki_E$. Data for adsorption of divalent cations could be fit by considering either charged or neutral complexes. In the current study with copper we only considered neutral complexes. In Eqs. (1) to (3) the concentrations and binding coefficients are given in units of M and $M^{-1}$, respectively. The program considers solution complexes, e.g., $(Cu^{2+}Cl^-)^+$ and $(Cu^{2+}Cl^-_2)^0$ and surface complexes, such as $(P^-Cu^{2+}Cl^-)^0$ or $(E^-Cu^{2+}Cl^-)^0$.

The concentration of the cation $M_j^{2+}$ at the surface is:

$$M_j^{2+}{}_{(0)} = S^{2+}Y_0{}^Z, \quad (4)$$

where $Y_0=\exp(-e\,\Psi_{(0)}/kT)$, Z is the valence of the given ion, $e$ is the absolute magnitude of an electronic charge, $\Psi_{(0)}$ is the surface potential, k is Boltzmann's factor, and T is the absolute temperature. $M_j^{2+}{}_{(0)}$ is the molar concentration of cation j close to the mineral layer, and $S^{2+}$ is the concentration of that cation in the equilibrium solution, far away from the surface.

For a cation of type i and valence z, the excess of its concentration in the double layer region is denoted by $Dl_i$ and given by:

$$Dl_i = Qz\frac{Si}{S^{(z)}} \quad (5)$$

$$\text{in which, } S^{(z)}=\Sigma S_i^{z+} \quad (6)$$

A priori only the total concentrations $C_i^z$ are known. For instance, for a divalent cation, j, in a solution with several monovalent and divalent anions, $A_k^-$, $A_m^{2-}$, the relation is:

$$C_j^{2+} = S_j^{2+} + DI_j + DI_{E,j} + (P^{2-}M^{2+})^0 + (E^{2-}M^{2+})^0 + \left(\sum_k (A_k^- S_j^{2+})^+ + ((A_k^-)_2 S_j^{2+})\right)^0 + (P^-(A_k^- M_j^{2+})^+)^0 + (E^-(A_k^- M_j^{2+})^+)^0) + \sum_m (A_m^{2-} S_j^{2+})^0 \quad (7)$$

The reduction in surface charge density is explicitly taken into account according to Equations (1) to (4). The computational procedure is an iterative process based on Equations (1) to (7), using the Gouy-Chapman equation for the surface charge density.

The binding coefficients of $Na^+$ and $K^+$ to montmorillonite were taken from Nir et al. (1986) and those of $Ca^{2+}$ and $Mg^{2+}$ were taken from Undabeytia et al. (1998). The association constants of the soluble complexes were taken from Lindsay (1979). Based on the results in Rytwo et al. (1996), all other association constants could be ignored in the range of cation concentrations (Cu, Ca, Mg) used in our study.

The required parameters were the binding coefficients $K_{2,Cu}$ and $H^+$ for the edge and planar sites, and those of $(CuCl)^+$.

The study was designed to enable a sequential determination of the parameters, as will be elaborated in the Results section. Each binding coefficient was determined by varying its value until the best fit was obtained between calculated and experimental results, the criteria being the maximization of $R^2$ and minimization of the root mean square error.

## Results and Discussion

Figure 1 shows that at low ionic strength the amount of Cu adsorbed increased with its added amounts, but at high ionic strength the adsorbed amounts reach a plateau at 24 mmol/kg clay. This outcome indicates that in the presence of high ionic strength the adsorption of Cu occurs mostly, or exclusively, on high affinity sites of montmorillonite, whose amount is 48 $mmol_c$/kg clay. We introduce an assumption that the high affinity sites for Cu adsorption are the edge sites of montmorillonite. Accordingly we performed model calculations in which we introduced a single binding coefficient, $K_{2E,Cu}$ (Eq. 3) for the adsorption of $Cu^{2+}$ on the edge sites. For the determination of this binding coefficient we considered all the data points in Figure 1 corresponding to the high ionic strengths, whereas for the low ionic strength only $Cu^{2+}$ concentrations below 24 μmol/g were considered, in order to avoid a contribution from $Cu^{2+}$ adsorption on the planar sites of the clay mineral. We deduced a value of $2x10^4$ $M^{-1}$ for $K_{2E, Cu}$. The high binding coefficient calculated for Cu adsorption on the edge sites suggests the formation of inner sphere complexes. In these calculations we assumed that the binding coefficients for $Ca^{2+}$, $Mg^{2+}$, $Na^+$ and $K^+$ binding to the edge sites were the same as previously used by ignoring site heterogeneity. Of course, this is a crude approximation, but it was adequate to explain the adsorbed amounts of all cations in the system. In the next stage we have determined the binding coefficient of $Cu^{2+}$ to the planar sites of montmorillonite, $K_{2, Cu}$, from additional 6 data points. The results also included predictions for the amounts of Ca and Mg remaining adsorbed. In these calculations we considered simultaneous adsorption of all the cations to both planar and edge sites. In the calculations we employed the binding coefficients from previous studies for adsorption of Ca, Mg and Na on the planar sites, and $K_{2E, Cu} = 2x10^4$ $M^{-1}$.

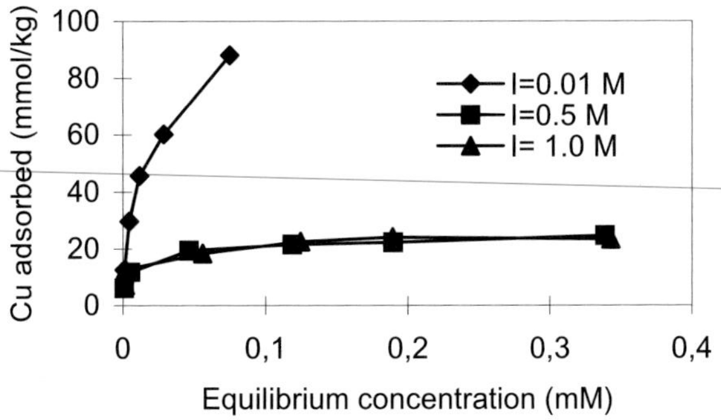


**Fig. 1.** Adsorption isotherms of Cu on montmorillonite at several ionic strengths (I) when using $NaClO_4$ as background electrolyte. Reprinted with permission from Undabeytia et al. (2002). Copyright 2002 American Chemical Society.

Table 1 shows that the model predictions are reasonably good for all the measured adsorbed amounts of the cations at the different ionic strengths used, by employing $K_{2, Cu} = 8$ $M^{-1}$ for Cu binding to the planar positions. At high total Cu concentrations, the decrease observed in the fraction of Cu adsorbed, upon increasing the ionic strength, is adequately explained in the model by a competition with sodium ions for occupancy of the planar sites, together with a decrease in the magnitude of the surface potential, which would reduce Cu binding and its accumulation in the double layer region. However, for the lowest total $Cu^{2+}$ concentrations the percentages of Cu adsorbed are almost independent of the ionic strength used. For these cases, our calculations indicate that $Cu^{2+}$ is almost exclusively adsorbed (≥ 98%) on the edge sites until reaching a value close to their saturation.

The set of binding coefficients which gave an overall best fit of pH with the experimental values were 10 $M^{-1}$ for $K_{1,H}$ and 18000 $M^{-1}$ for $K_{1E,H}$. Ignoring proton binding produces only a slight variation in the calculated amount of Cu adsorbed (<1.2%), which is due to the much smaller initial amount of $H^+$ in comparison to the other cations present in the system.

**Table 1.** Percentages of Cu, Ca and Mg adsorbed on montmorillonite as a function of the ionic strength and the amount of Cu added. Experimental and calculated values. The background electrolyte used was $NaClO_4$. [a,b] Reproduced with permission from Undabeytia et al. (2002). Copyright 2002 American Chemical Society.

| Ionic strength (M) | Cu added (μM) | Cu (%) | | Ca (%) | | Mg (%) | |
|---|---|---|---|---|---|---|---|
| | | Exp. | Calc. | Exp. | Calc. | Exp. | Calc. |
| 0.01 | 31.0 | 99.3 | 99.7 | 86.7 | 83.7 | 84.2 | 81.0 |
| | 63.7 | 98.4 | 99.5 | 86.2 | 83.0 | 83.4 | 80.3 |
| | 152.0 | 97.1 | 94.2 | 84.7 | 81.3 | 81.6 | 78.4 |
| | 239.0 | 95.1 | 87.9 | 83.0 | 78.8 | 80.0 | 76.8 |
| | 329.0 | 91.3 | 83.9 | 81.6 | 78.4 | 78.2 | 75.2 |
| | 516.0 | 85.4 | 78.7 | 78.4 | 75.4 | 74.3 | 72.0 |
| 0.5 | 31.5 | 97.6 | 94.2 | 13.4 | 10.2 | 11.6 | 8.5 |
| | 63.9 | 91.8 | 90.2 | 15.3 | 9.9 | 11.2 | 8.2 |
| | 143.0 | 74.1 | 68.4 | 15.3 | 9.6 | 11.4 | 7.9 |
| | 226.0 | 47.6 | 50.0 | 15.8 | 9.5 | 10.9 | 7.9 |
| | 300.0 | 37.0 | 40.7 | 15.7 | 9.5 | 10.3 | 7.9 |
| | 461.0 | 29.9 | 35.1 | 15.5 | 9.4 | 9.4 | 7.8 |

a. The calculations employed the following values of binding coefficients for planar sites: Cu (8 $M^{-1}$) (current work); Ca (10 $M^{-1}$), Mg (8 $M^{-1}$) as in Undabeytia et al. (1998) and Rytwo et al. (1996); Na (1 $M^{-1}$); H (10 $M^{-1}$) (current work) and K (4 $M^{-1}$) as in Nir et al. (1986). For the edge sites the same binding coefficients were used, except for copper and the proton, for which a binding coefficient of $2 \times 10^4$ $M^{-1}$ and $1.8 \times 10^4$ $M^{-1}$ , respectively, were used. The total concentrations in the system (clay + solution) were 2.52 mM for Ca, 0.479 mM for Mg, and 0.048 mM for K (Na concentration varied with the ionic strength used).

b. The relative standard deviations obtained for Cu adsorption ranged between 0.1 and 5.0 %.

The introduction of the chloride anion instead of the perchlorate anion reduces the amount of Cu adsorbed on the clay, due to the complexation reaction between copper and chloride (Table 2). The binding coefficient of $CuCl^+$ was determined to be 20 $M^{-1}$ for both the planar and edge sites. The calculated adsorbed amounts of total Cu agree reasonably well with the experimental results. Table 2 shows that for the lowest ionic strength the fraction adsorbed as the monovalent cation $CuCl^+$ is negligible, becoming relatively important only for the highest ionic strength.

**Tab. 2.** *Percentages of Cu adsorbed on montmorillonite sites and its speciation in the adsorption experiments and after three desorption cycles as a function of ionic strength and the amount of Cu added. The background electrolyte used was NaCl. The calculated and measured pH is also shown.*† *In two cases the results comprise Cu adsorption at equilibrium pH 3.5.*

| Ionic strength (M) | Cu added (μM) | Step | Cu adsorbed (%)‡ | | Cu adsorbed (%)§ | | | Equilibrium pH | |
|---|---|---|---|---|---|---|---|---|---|
| | | | | | $Cu^{2+}$ | | ***Total*** | | |
| | | | Exp | Calc | planar | edge | $CuCl^+$ | Calc | Exp |
| 0.01 | 81.1 | Ads. | 95.3 | 99.2 | 3.1 | 96.9 | - | 6.70 | 7.00 |
| | | Des. #3 | 96.9 | 99.7 | 2.6 | 97.4 | - | 7.07 | 6.86 |
| 0.01 | 81.1 | Ads. | 87.0 | 79.6 | 75.0 | 24.6 | 0.4 | 3.24 | 3.67 |
| 0.01 | 161.2 | Ads. | 91.6 | 93.2 | 25.6 | 74.4 | 0.1 | 5.86 | 6.11 |
| | | Des. #3 | 96.0 | 97.2 | 21.5 | 78.5 | - | 6.51 | 6.39 |
| 0.01 | 489.1 | Ads. | 79.9 | 78.9 | 69.0 | 30.5 | 0.5 | 5.55 | 5.76 |
| | | Des. #3 | 92.7 | 90.5 | 66.8 | 32.9 | 0.3 | 6.18 | 6.10 |
| 0.01 | 489.1 | Ads. | 65.8 | 71.8 | 84.4 | 15.1 | 0.4 | 3.29 | 3.61 |
| 0.1 | 75.2 | Ads. | 91.4 | 96.2 | 1.8 | 97.9 | 0.3 | 6.50 | 6.59 |
| | | Des. #3 | 98.3 | 98.6 | 1.2 | 98.7 | 0.1 | 6.91 | 6.73 |
| 0.1 | 487.2 | Ads. | 50.1 | 49.0 | 44.4 | 48.0 | 7.2 | 5.55 | 6.00 |
| | | Des. #3 | 79.2 | 72.1 | 32.2 | 64.0 | 3.8 | 6.37 | 6.19 |

§. Percent adsorbed of total Cu adsorbed.
†. The percentages are expressed with respect to the total amounts (adsorbed + in solution) for each step.
‡. The relative standard deviations obtained for Cu adsorption ranged between 0.1 and 2.3%.

The effect of a pH decrease on Cu adsorption is shown in Figure 2, where the equilibrium pH was adjusted to 3.5, whereas that of the uncontrolled solutions varied between 6.11 and 7.00. The speciation in solution and the adsorbed amounts of some of the points of the isotherm at pH 3.5 are included in Table 2.

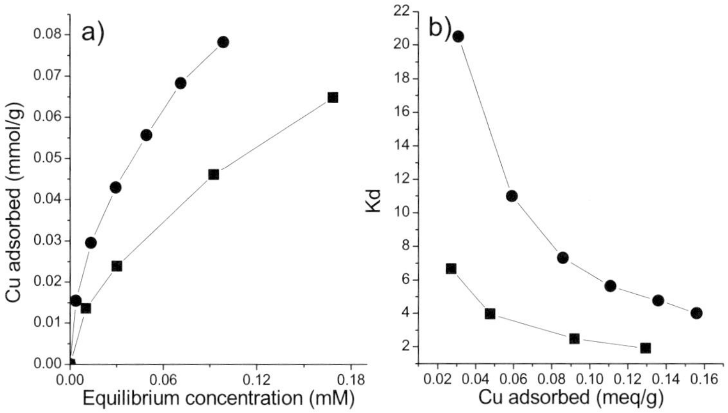


**Fig. 2.** Adsorption isotherms of Cu (a) as well as their Kd values as a function of Cu loading (b) at equilibrium pH free (O) and 3.5 (■).

The amount of Cu adsorbed decreases at the lower pH, and is reflected in much lower values of the distribution coefficient Kd, which is a classical method to express the adsorbent-adsorbate affinity. High values of Kd at low surface coverage and their decrease at larger loading are attributed to the performance of high and low affinity sites at low and high loading, respectively. Near saturation of the sites of high affinity, the adsorption on those of lower affinity emerges, which would explain the decrease in Kd values (Maqueda et al. 1998).The ability of the model to yield reasonably good predictions of the experimental results at low pH further supports the suggestion that the high affinity sites are due to edge sites on which Cu adsorption decreases largely because of the competition with $H^+$. For Cu sorbed on planar positions the competition with the proton is not so strong, because of the similar and low values of the binding coefficients. Consequently, the decrease in pH would produce a more drastic decrease in the adsorption of Cu at low rather than at a high loading. This study on the pH effect on Cu sorption was firstly designed to try to make an assignment of the high and low affinity sites as a function of the loading reported in the literature, and secondly, to present additional experimental results of adsorption and test the ability of the model to predict cases not included previously in the determination of the binding coefficients. The model proved to be a successful tool for the analysis of these experimental results.

Table 2 also illustrates that the calculated fractions of Cu adsorbed on the planar sites decrease throughout the desorption cycles. This effect is more pronounced in the case of the highest total Cu concentration used (487.2 μM) at 0.1 M NaCl. Our calculations indicate that there is almost no desorption from the edge sites for the highest Cu concentration used, whereas the desorption from the edges sites is very small for the lowest Cu concentration used.

The hysteresis arising from planar sites as observed for the highest Cu concentration added in 0.01 M NaCl can be mostly explained by the fact that the desorption cycles involve centrifugation, removal of the half of the supernatant and addition of a corresponding volume of 10 mM NaCl, and thus the total concentrations of Ca and Mg which affect Cu adsorption due to competition are reduced. Part of this effect is due to a small increase in the magnitude of the surface potential due to a smaller sum of total concentrations of divalent cations in desorption cycles, which results in higher $Cu^{2+}$ or $CuCl^{+}$ concentrations at the surface according to Eq. (4). Indeed, when supplementing the amounts of Ca and Mg desorbed in the adsorption stage for the first desorption step the amount of Cu desorbed increases and the hysteresis is largely reduced (data not shown).

Adsorption of the pesticide chlordimeform on montmorillonite is strongly dependent on the state of aggregation of the clay platelets (Undabeytia et al. 1999). The shape of the isotherm changes from that of a low affinity (S-type) on Ca-montmorillonite to one of high affinity (L-type) on Na-montmorillonite (Figure 3). For Ca-montmorillonite the clay particles are associated through face-to-face aggregation unlike Na-montmorillonite for which the clay platelets are largely dispersed in solution. Consequently, the interaction of the pesticide with the clay particles will be less facilitated on Ca-montmorillonite, resulting in its decreased adsorption. The affinity of chordimeform for this clay increases after some molecules are adsorbed because of an increase of the distances between closely apposed clay platelets. This is the reason for the need to use a lower binding coefficient in the modeling of the adsorbed amounts of the pesticide for the lowest concentrations used (30 $M^{-1}$), in contrast to a higher binding coefficient (90 $M^{-1}$) for the rest of concentrations used (Table 3).

The state of aggregation of the clay platelets also plays a role in the hysteresis observed. When the concentrations of $Ca^{2+}$ and $Mg^{2+}$ are supplemented in the desorption processes from Ca-montmorillonite as in Cu desorption, the hysteresis is not completely abolished. The remaining hysteresis can be anticipated from the S-shape of the adsorption curve, which implies a larger affinity of chlordimeform for adsorption to Ca-montmorillonite at larger chlordimeform concentrations. The Ca-montmorillonite-chlordimeform system yields a higher affinity for chlordimeform adsorbed despite the lower chlordimeform concentrations in the desorption steps than in the adsorption stage, where the added chlordimeform encounters more clay platelets in aggregates that pose more steric inhibition for chlordimeform adsorption.

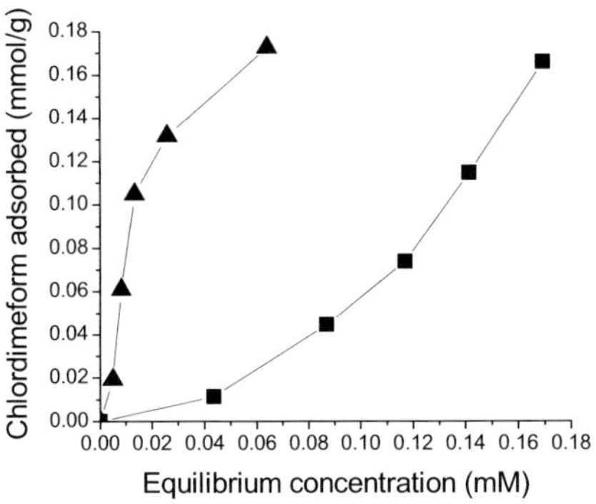


**Fig. 3.** Adsorption isotherms of chlordimeform on Na (▲)- and Ca (■)-montmorillonite. The background electrolyte was 0.01 M NaCl. Reprinted with permission from Undabeytia et al. (1999). Copyright 1999. American Chemical Society.

The influence of the presence of the pesticide chlordimeform on Cu adsorption is shown in Table 3. The model is able to yield good predictions of the adsorbed amounts of the pesticide and the metal. In the simultaneous adsorption of both cations, Cu adsorption is hardly affected by the presence of the pesticide at the lowest Cu concentration added, since the adsorption occurs almost exclusively on edge sites; consequently the fraction of Cu adsorbed is close to unity. At higher Cu concentrations, a 10-fold increase in the added concentration of chlordimeform results in a 8% decrease in the adsorbed amounts of the metal. In this case, the calculated amounts of Cu adsorbed indicate that the fraction of Cu adsorbed on the planar positions decreases, unlike that on the edge sites, i.e., the pesticide is competing effectively with Cu for adsorption on the planar positions. Conversely, a 6-fold increase in the Cu concentration gives a 7.7% reduction in the adsorption of chlordimeform at the lowest concentration used, but the reduction only amounts to 3.6% for higher chlordimeform concentrations. This reduction is due to the higher affinity of chlordimeform to the clay sites at higher loading, as deduced from the S-shape adsorption isotherm.

**Tab. 3.** *Percentages of Cu and chlordimeform adsorbed on montmorillonite sites and Cu speciation in adsorption experiments and after three desorption cycles as a function of the amount of Cu and chlordimeform (Chl.) added. Experiments include simultaneous addition of Cu and Chl. (sim.), or Cu addition after treating the clay mineral with Chl. (suc.). The background electrolyte used was 0.01 M NaCl.*

| Case | Chl. added (μM) | Cu added (μM) | Step | Cu adsorbed (%)‡ | | Cu adsorbed (%)§ Cu²⁺ | | *Total* | Chl. adsorbed (%) | |
|---|---|---|---|---|---|---|---|---|---|---|
| | | | | Exp | Calc | planar | edge | $CuCl^+$ | Exp | Calc |
| Sim. | 101 | 79 | Ads. | 96.5 | 99.2 | 3.0 | 97.0 | <0.1 | 56.8 | 54.9 |
| | | | Des.#3 | 98.3 | 99.7 | 2.5 | 97.5 | <0.1 | 62.3 | 63.5 |
| | 105 | 482 | Ads. | 80.3 | 78.8 | 68.9 | 30.8 | 0.3 | 49.1 | 49.6 |
| | | | Des.#3 | 93.2 | 90.0 | 66.2 | 33.6 | 0.2 | 57.4 | 60.8 |
| | 517 | 80 | Ads. | 95.9 | 99.0 | 3.1 | 96.9 | <0.1 | 77.7 | 75.2 |
| | 514 | 479 | Des.#3 | 77.1 | 76.0 | 67.7 | 32.0 | 0.3 | 72.6 | 71.6 |
| | 1040 | 80 | Ads. | 95.0 | 98.7 | 3.1 | 97.0 | <0.1 | 83.4 | 72.5 |
| | | | Des.#3 | 97.9 | 99.5 | 2.5 | 97.5 | 0 | 89.8 | 80.7 |
| | 1023 | 483 | Ads. | 72.3 | 72.7 | 66.5 | 33.1 | 0.4 | 79.8 | 69.2 |
| | | | Des.#3 | 92.5 | 88.3 | 63.2 | 36.6 | 0.2 | 88.5 | 79.6 |
| Suc. | 100 | 73 | Ads. | 98.5 | 99.7 | 2.49 | 97.5 | <0.1 | 67.4 | 63.5 |
| | 100 | 462 | Ads. | 82.9 | 85.6 | 70.3 | 29.5 | 0.3 | 61.6 | 56.3 |
| | 1000 | 73 | Ads. | 98.2 | 99.6 | 2.5 | 97.5 | <0.1 | 88.5 | 81.2 |
| | 1000 | 463 | Ads. | 81.3 | 83.5 | 69.5 | 30.2 | 0.3 | 85.4 | 76.6 |

§. Percent adsorbed of total Cu adsorbed.

†. The percentages are expressed with respect to the total amounts (adsorbed + in solution) for that step.

‡. The relative standard deviations obtained for Cu adsorption ranged between 0.1 and 2.3%.

Tables 2 and 3 indicate that Cu adsorption is slightly higher when the clay was previously treated with the pesticide than in the case of adsorption from pesticide-free solutions or when both adsorbates were added simultaneously. This may be due to the fact that adsorption of the pesticide on the planar positions of the clay promotes an opening between the platelet layers, thus facilitating the subsequent penetration of Cu. The loading of the pesticide on the clay is very small, so that only a small fraction of the interlamellar positions are occupied by the pesticide, the other positions being available for metal sorption. This behaviour is more evident at the highest Cu concentrations, where adsorption of the metal on the planar positions is larger, and accordingly, the fraction of Cu adsorbed on the planar positions increases with respect to that of Cu adsorption from pesticide-free solutions or when Cu is simultaneously added with the pesticide.

## Acknowledgements

The financial support of the Spanish Comisión Interministerial de Ciencia y Tecnología (CICYT), under project AGL2002-00993, and Junta de Andalucía (PAI RNM166) is acknowledged herewith. Prof. Shlomo Nir acknowledges the Spanish Government for a sabbatical stay at IRNAS (ref. SAB2003-0190).

## References

Lindsay WL (1979) Chemical Equilibria in Soils. Wiley New York.

Maqueda C, Undabeytia T, Morillo E (1998) Retention and release of copper on montmorillonite as affected by the presence of a pesticide. J Agric Food Chem 46:1200-1204

Nir S (1986) Specific and nonspecific cation adsorption to clays: Solution concentrations and surface potentials. Soil Sci Soc Am J 50:52-57

Nir S, Hirsch D, Navrot J, Banin A (1986) Specific adsorption of lithium, sodium, potassium and strontium to montmorillonite: Observations and predictions. Soil Sci Soc Am J 50:40-45

Rytwo G, Banin A, Nir S (1996) Exchange reactions in the Ca-Mg-Na-montmorillonite system. Clays Clay Min. 44:276-285

Undabeytia T, Morillo E, Maqueda C (1996) Adsorption of Cd and Zn on montmorillonite in the presence of a cationic pesticide. Clay Min. 31:485-490

Undabeytia T, Nir S, Rytwo G, Morillo E, Maqueda C (1998) Modeling adsorption-desorption processes of Cd on montmorillonite. Clays Clay Min. 46:423-428

Undabeytia T, Nir S, Polubesova T, Rytwo G, Morillo E, Maqueda C (1999) Adsorption-desorption of chlordimeform on montmorillonite: Effect of clay aggregation and competitive adsorption with cadmium. Environ Sci Technol 33:864-869

Undabeytia T, Nir S, Rytwo G, Serban C, Morillo E, Maqueda C (2002) Modeling adsorption-desorption processes of Cu on edge and planar sites of montmorillonite. Environ Sci Technol 36:2677-2683

Zachara JM, Smith SC, Resch CT, Cowen CE (1992) Cadmium sorption to soil separates containing layer silicates and iron and aluminum oxides. Soil Sci Soc Am J 56:1074-10784

Zachara JM, McKinley JP (1993) Influence of hydrolysis on the sorption of metal cations by smectites: Importance of edge coordination reactions. Aquatic Sci 55:250-261

# NMR spectroscopy: a tool to study interactions between organic pollutants and soil components?

A.M. Delort, B. Combourieu, N. Haroune, P. Besse, M. Sancelme

Dept. of Chemistry, Laboratoire Synthèse et Etude de Systèmes à Intérêt Biologique, UMR 6504 CNRS- University Blaise Pascal, Aubière, France

Corresponding author: Laboratoire Synthèse et Etude de Systèmes à Intérêt Biologique, UMR 6504 CNRS- University Blaise Pascal, 63117 Aubière cedex, France
– Tel: +33 473 40 77 14 – fax: +33 473 40 77 17 – e-mail: A-Marie.DELORT@ univ-bpclermont.fr

## Abstract

This paper reports studies from selected recent papers using NMR to investigate interactions between organic pollutants and soil matrices including SOM fractions, clays, whole soils or sediments, three main approaches are presented. First, an indirect approach consists to measure organic carbon normalized sorption coefficients ($K_{oc}$ values) of organic pollutants with soil matrices and then to establish correlations between these values and structural elements of soil organic matter (carboxylic, carbonyl, aromatic, aliphatic groups…). In that case, NMR technique only plays a role in determining soil organic matter structure. Second are reported solid state NMR ($^{13}$C and $^{15}$N CP-MAS, cross polarization magic angle spinning) experiments which are the main tools to investigate covalent bonds between organic pollutants and SOM. Finally are described experiments carried out on hydrated solid matrices using HR-MAS (high resolution magic angle spinning) NMR. This approach, which is the most recent and novel, allows characterizing molecular species at the solid-aqueous interface, these species are differentiated according to their degree of mobility. It should help to investigate the major question of bio-availability of organic pollutants, mainly for bacteria.

## Keywords

NMR; HR-MAS; CP-MAS; covalent bonds; non-covalent bonds; hydrated matrices; SOM; soil components

## Introduction

The fate of organic pollutants in soils from the surface to ground waters is greatly dependent on the interactions they can make with soil components and on degradation processes, mainly biodegradation by microorganisms. NMR spectroscopy which is the major tool for chemists to analyze more or less pure organic molecule structures, has been used more recently directly in situ on complex media, without purification, leading to applications in the field of environment (Grivet et al. 2003). Many studies are related to microbial degradation of xenobiotics; the identification of intermediary and final metabolites allows to describe metabolic pathways in detail (for review see Delort and Combourieu 2001, 2000; Combourieu et al. 2003; Grivet et al. 2003). NMR is also a tool to study interactions between organic pollutants and soil components, including humic and fulvic acids (HA and FA), humin, clays… When covalent bonds are created, these pollutants are more persistent; they can participate to the humification process or induce some eco-toxicological problems. On the contrary these xenobiotics can interact without forming covalent bonds; in that case they are likely to be more bio-available for microbes or plants or to be transported to ground water. Although reported studies were carried out exclusively under laboratory and not under field conditions because of the relatively low sensitivity of NMR techniques, this approach remains one of the unique tool to assess interaction mechanisms at a molecular level. In this paper we shall present selected examples illustrating three NMR approaches developed to study organic pollutant interactions with soils. These case studies were chosen almost exclusively from recent papers (last five years). First, an indirect approach consists to measure organic carbon normalized sorption coefficients ($K_{oc}$ values) of organic pollutants with soil matrices and then to establish correlations between these values and structural functions of soil organic matter (carboxylic, carbonyl, aromatic, aliphatic groups...). In that case, NMR technique only plays a role in determining soil organic matter structure (SOM). Second are reported solid state NMR experiments ($^{13}C$ and $^{15}N$ CP-MAS, cross polarization magic angle spinning) which are the main tools to investigate covalent bonds between organic pollutants and SOM. Finally are described NMR experiments designed to assess non covalent interactions. Apart from liquid state NMR experiments (chemical shift or T1 measurements) carried out in model systems, HR-MAS (high resolution magic angle spinning) NMR technique applied on hydrated solid matrices is described in detail. This latter approach, which is the most recent and novel, allows characterizing molecular species at the solid-aqueous interface, these species being differentiated according to their degree of mobility. It should help to investigate the major question of bio-availability of organic pollutants, mainly for bacteria.

For a more extensive literature, readers can refer to an interesting book published about this topic (Nanny et al. 1998) and a recent review by our group (Delort et al. 2004). Although some papers report interactions between organic pollutants and inorganic fractions of soil, most studies are focused on interactions with SOM.

## Correlations between $K_{oc}$ and soil organic matter structure

NMR represents a major technique to investigate SOM structure; in particular to describe the main functional groups (carboxylic, carbonyl, aromatic, aliphatic groups...). It is a non destructive approach to analyze samples, including soils and sediments, compost, peat, HA and FA. $^{13}C$ CP-MAS (cross polarization magic angle spinning) NMR has been extensively used to study solid matrices in a solid state (for review see Kögel-Knabner 2000, Preston 2001, Keeler and Maciel 2003). More recently, innovative approaches such as multidimensional $^{1}H$ NMR experiments or $^{1}H$ spin-spin relaxation time ($^{1}H$-$T_2$) measurements have brought some new insight in SOM structure (Simpson 2001, Simpson et al. 2002, 2003b, Gunasekara et al. 2003, Cook et al. 2003). To investigate interactions between organic pollutants and soil matrices, it is possible to establish correlations between organic carbon normalized sorption coefficients ($K_{oc}$ values) and structural elements of SOM. $K_{oc}$ values are measured in solution with pollutants at natural abundance, under relatively low concentrations. Recent papers report studies of PAHs (Xing 2001, Salloum et al. 2002, Gunasekara et al. 2003, Simpson et al. 2003a, Kang and Xing 2005), trinitrotoluene and its metabolites (Eriksson et al. 2004), pesticides (Ahmad et al. 2001, Kulikova and Perminova 2002), volatile organic compounds (Kile et al. 1999, Shih and Wu 2002).

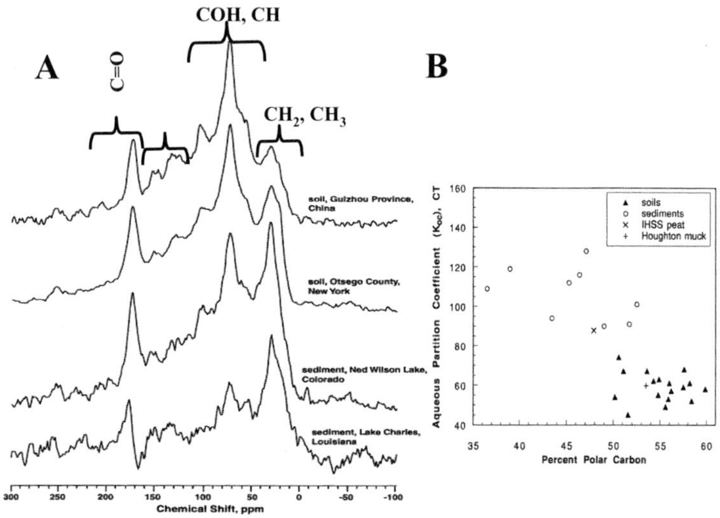


**Fig. 1.** Correlations between $K_{oc}$ and soil organic matter structure. A) $^{13}$C-CP-MAS NMR spectra of selected soils and sediments. Percent polar carbon (POC) was calculated as the combined fractions of 50-105 ppm and 160-220 ppm regions of the spectra. B) Plot of $K_{oc}$ vs POC for soils and sediments. An inverse correlation between partition efficiency of carbon tetrachloride (CT) to SOM and SOM polarity is observed. (adapted from Kile et al. 1999)

Figure 1 shows the application of this approach to the case study of the interaction of carbon tetrachloride (CT) with various soils and sediments (Kile et al. 1999). In figure 1A are reported $^{13}$C-CP-MAS spectra of four selected soils and sediments. Characteristic functional group carbons can be assigned to four different spectral regions including aliphatic carbons (δ 0-50 ppm), alcoholic carbons in carbohydrates, ethers, anomeric carbons (δ 50-150 ppm ) aromatic and phenolic carbons (δ 105-160 ppm), carboxyl, carbonyl carbons in aldehydes, amides and ketones (δ 160-220 ppm). Percent polar carbon (POC) was calculated as the combined fraction of the 50-105 and 160-220 ppm regions of theses $^{13}$C spectra. In figure 1B the plot of $K_{oc}$ of Ct vs the percent of POC for various soils and sediments shows a clear inverse trend between $K_{oc}$ and POC, with two distinct populations of soils and sediments. This result strongly suggests that the partition efficiency of CT, and more generally of non-polar compounds, to SOM is dependent on the polarity of SOM.

Although this approach is indirect, it could be used in principle as a predictive tool to extrapolate possible interactions between a pollutant and a soil of known composition. Also, both soil and pollutants can be studied at natural abundance which is a great advantage.

## Covalent bonds with organic matter in solid state matrices

$^{13}$C-CP-MAS or $^{15}$N-CP-MAS have been used to study covalent interactions between organic pollutants and soil organic matter resulting from soil microbial activity. Because studies were carried out under laboratory conditions; these covalent bonds had to be created by incubating organic pollutants in the presence of soil or soil components with enzymes or microorganisms; or by making artificial compost. For these studies, $^{15}$N or $^{13}$C enriched xenobiotics were used because of the low sensitivity of NMR and the low concentration of organic pollutants compared to the amount of soil organic matter. Isotopic enrichment is needed to detect specifically residues resulting from covalent bonding. Indeed, SOM being highly concentrated, significant $^{13}$C or $^{15}$N NMR signals can be detected even at natural abundance. To increase even more the isotopic enrichment difference between the organic pollutant and SOM, some authors have suppressed the natural abundance enrichment of SOM by preparing $^{13}$C- or $^{15}$N –depleted compost (Berns et al. 2005, Haider et al. 1993). For that, they have used plants grown under $^{12}CO_2$ or $^{14}NH_4$ conditions. As labeled pollutants are not commercially available but must be synthesized in specialized laboratories, only a few types of compounds have been studied up to date. Recent papers describe the interactions with $^{15}$N-TNT (Achtnich et al. 1999, Bruns-Nagel et al. 2000, Knicker et al. 1999, 2001, 2002, Knicker 2003, Thorn et al. 2002, Thorn and Kennedy 2002, Thiele et al. 2002), $^{13}$C-$CH_3Br$ (Tao and Maciel 2002), $^{13}$C-PAHs (Käcker et al. 2002), $^{13}$C-benzothiazoles (Witte et al. 1998, 2002), $^{13}$C-atrazine (Benoit and Preston 2000), $^{15}$N-simazine (Berns et al. 2005). One of the problems of this approach is that the interpretation of NMR spectra is not always straightforward because organic pollutants are usually labeled at a single position. As a result, NMR only gives partial information about the molecular structure. Also it is sometimes very difficult to assume that observed signals result from covalent bonding and not from sequestration. Therefore additional experiments to NMR must be carried out to demonstrate the presence of bound residues: for instance, chemical cleavage of covalent bonds followed by mass spectroscopy analysis (Käcker et al. 2002) or silylation of the bound residues followed by size exclusion chromatography analysis (Berns et al. 2005). To partially overcome these problems, Knicker (2003) recently proposed a new approach, which combines the use of $^{15}$N-TNT and $^{13}$C-labelled plant material with $^{15}$N-$^{13}$C DCP-MAS (double cross polarization magic angle spinning) NMR experiments.

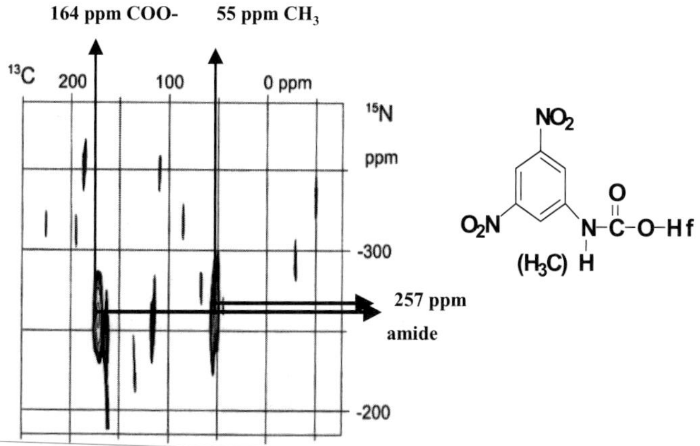


**Fig. 2.** Two-dimensional solid-state DCPMAS $^{15}$N $^{13}$C NMR spectrum of the acid insoluble fraction obtained from $^{13}$C-enriched plant residues that were incubated after addition of $^{15}N_3$-TNT for 11 months. Cross peaks demonstrate the formation of amide linkages and alkylation of their N. Hf: humic fraction. (adapted from Knicker 2003).

An example of $^{15}$N-$^{13}$C DCP-MAS NMR spectrum collected from the acid-insoluble fraction obtained from $^{13}$C-enriched plant residues that were incubated after addition of $^{15}N_3$-TNT for 11 months is presented in figure 2. Two large $^{13}$C-$^{15}$N cross peaks correspond to the correlations between nitrogen from amide groups ($^{15}$N $\delta = 257$ ppm) with carbon from carboxylic groups ($^{13}$C $\delta = 164$ ppm) and carbon from methyl groups ($^{13}$C $\delta = 55$ ppm) respectively. This result proves unambiguously the formation of amide linkages between N of TNT and carboxylic groups of this humic fraction; in addition it shows that most of the N of these amide linkages are alkylated. This technique is extremely powerful and novel but remains limited by its low sensitivity: some experiments took up to 48 hours, and some fractions could no be analyzed due to their low $^{15}$N concentrations.

Finally, one should remind that, in most cases, CP-MAS NMR is not absolutely quantitative due to the differential magnetization transfer efficiency between $^{13}$C or $^{15}$N nuclei and protons, depending on the chemical structure of the molecules. Preston (2001) wrote a review about technical problems encountered with the use of this technique for studying SOM. Also structural information can be limited due to a lower resolution of solid state NMR spectra (broad lines). In spite of these various drawbacks, $^{13}C$ and $^{15}N$ CP-MAS NMR experiments remain a method of choice to study organic pollutants/ SOM covalent interactions because this technique is non destructive and soil samples can be directly studied in the NMR tube.

## Mobility of pollutants at the solid-aqueous interface in hydrated matrices.

Weak interactions are usually difficult to analyze. One can take advantage of NMR properties connected with the orientation or movements of the molecules to assess these interactions.

A first approach is based on studies with model systems in liquid solution (for instance pollutant/ extracted humic acids…). A basic technique is to look at NMR shifts which can give indications about the type of interaction involved ($\pi$–$\pi$, H bonds…) (Zhu et al 2004). Also, many authors have used spin-lattice relaxation time ($T_1$) values to describe molecular interactions in solution (for review see Simpson et al. 2004b, Delort et al. 2004, Nanny et al. 1998). $T_1$ is basically an NMR parameter reflecting the mobility of each atom as it is inversely proportional to the correlation time $\tau_c$ Consequently any decrease in the contaminant $T_1$ value is attributable to a molecular association with the soil component.

Although these studies are of great interest, experimental conditions (liquid state) are rather far from those encountered in soils. Hydrated solid matrices (peat, soil, clay) can be considered as more relevant to real environmental conditions. The challenge under these conditions is to study adsorption-desorption mechanisms at the solid-aqueous interface. These mechanisms influence the mobility of organic compounds at the surface of a colloid. It is now possible to assess this mobility of pollutants at the solid-aqueous interface in hydrated matrices with the recent developments of HR-MAS (high resolution magic angle spinning) technique combined to specific HR-MAS probes allowing the use of gradients and shimming coils. Basically, this technique allows the characterization by NMR of inhomogeneous compounds with liquid-like dynamics (Piotto et al. 2001). In hydrated soil matrices, very mobile species give narrow signals, close to liquid state NMR spectra, on the contrary when the strength of the interaction increases the signals are more and more broadened. Therefore NMR signals reflect the degree of mobility of pollutants at the liquid/solid interface. Note that this technique, like CP-MAS, is non destructive for solid samples. Various nuclei have been used, including $^{19}F$, $^{1}H$ and $^{31}P$; these nuclei are very sensitive and can be easily detected at natural abundance. Khol et al. (2000) studied the sorption of $^{19}F$-hexafluorobenzene ($^{19}F$-HFB) on peat samples hydrated up to 65% of the maximum $H_20$ capacity. On $^{19}F$-MAS NMR spectra narrow lines were assigned to free, liquid HFB and lousy bound, mobile HFB; on the contrary a broad line was due to immobile HFB (Figure 3).The evolution of these different species with time was clearly observed; after 24 hours all the free HFB was adsorbed on peat. Complementary experiments showed that at least three sorption sites of HFB could be identified on molecular weight fractions of soil humic acid (Khalaf et al. 2003).

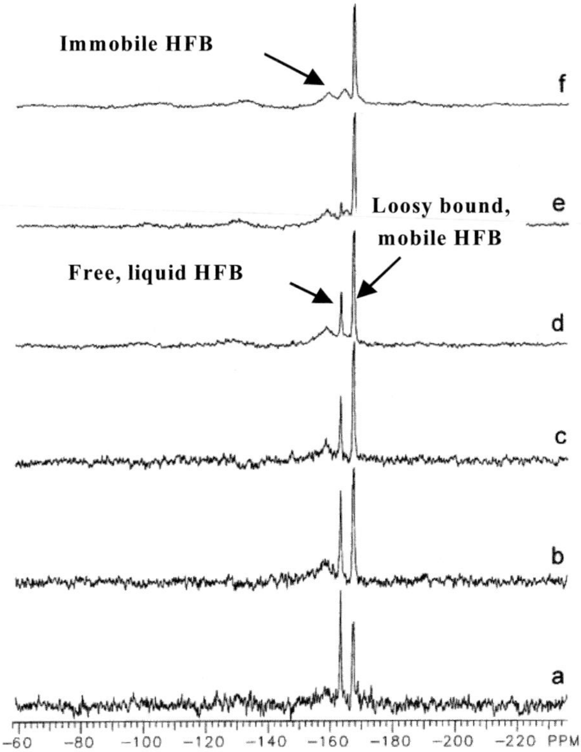


**Fig. 3.** $^{19}F$ solid-state NMR spectra of HFB sorbed to whole peat wetted to 65% of the maximum $H_20$ capacity. Spectra were aquired during 17 min (a), 25 min (b), 37 min (c), 1h (d), 4h (e) and 24h (f) after application of HFB. Three HFB signals could be detected corresponding to free liquid HFB, loosy bound, mobile HFB and immobile HFB. (adapted from Khol et al. 2000).

$^1$H HR-MAS was applied to study the interaction of pesticides with soil (Simpson et al. 2001) or clay (Combourieu et al. 2001). Opposite to $^{19}$F NMR which is easy to apply because there is no problem of overlapping signals due to the matrix and the very large chemical shift range of $^{19}$F nuclei (about 500 ppm), some difficulties are encountered with $^1$H NMR: the narrow chemical shift range of $^1$H nuclei (about 15 ppm), the presence of a huge signal due to water in the samples, or the numerous $^1$H resonances due to SOM. However, these problems can be overcome by applying multidimensional $^1$H NMR and water suppression techniques. The main advantage of $^1$H NMR is the ubiquitous presence of protons in organic pollutants at natural abundance and the highest sensitivity of detection of this nucleus. We have demonstrated the potential of the $^1$H HR-MAS NMR technique to study interactions of the herbicide MCPA (4-chloro-2-methylphenoxyacetic acid) with highly hydrated anionic clays, namely layered double hydroxides (LDHs) (Combourieu et al. 2001). By using a well-characterized model of soil it was possible to distinguish unambiguously the mobile (adsorbed) and immobile (intercalated) pesticide. Figure 4 shows the $^1$H signals of MCPA adsorbed at the surface of 200% $D_2O$ hydrated clay ($Mg_3AlCl$), sharp signals are characteristic of weakly bound compounds (top trace). On the contrary, $^1$H from intercalated MCPA in the clay layers are not visible being broadened, this results from very strong interactions with the matrix (bottom trace).

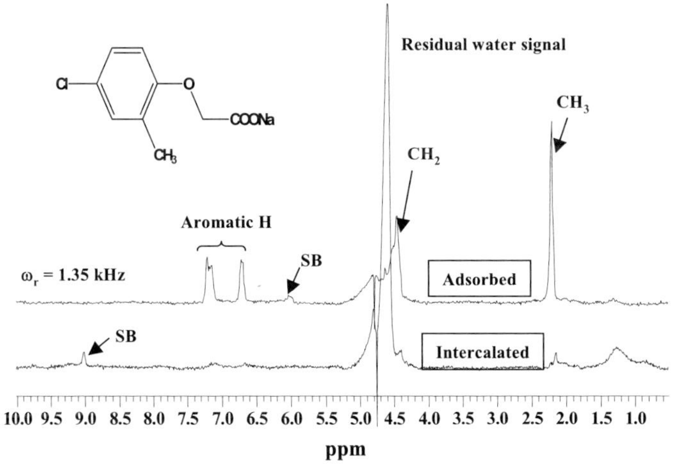


**Fig. 4.** $^1$H HR-MAS NMR spectra of adsorbed (top trace) and intercalated (bottom trace) MCPA clay ($Mg_3AlCl$) hydrated with $D_2O$. Sharp signals of adsorbed MCPA indicate an efficient tumbling of the molecules at the surface, while signals from MCPA tightly packed inside the structure are not observable by HR-MAS. r: spinning rate of the rotor. SB: spinning side bands.

Simpson et al. (2001) could observe directly interactions between the herbicide trifluarin and the surrounding soil matrix, swollen in water. In a recent work, we have studied the adsorption of a pesticide named glyphosate ((N-phosphonomethyl)glycine) at the surface of LDHs through different HR-MAS NMR approaches. In addition to $^{1}H$ NMR, $^{31}P$ was used as a probe to appreciate the surface distribution of the pesticide since $^{31}P$ chemical shifts are very sensitive to surrounding lattice, including local pH. Three types of glyphosate molecular species could be distinguished and characterized (unpublished data).

## Conclusion

Understanding of molecular mechanisms involved in the interactions between organic pollutants and soil is extremely difficult. Investigations are limited by the lack of adequate tools to assess these questions. NMR spectroscopy, although it is not very sensitive, has become a promising approach. Beside classical studies by $^{13}C$ CP MAS NMR experiments, new methods allow deeper insight in molecular structure of SOM. A recent paper describes procedures for the generation of 2D NMR databases containing spectra predicted from chemical structures. Using spectral prediction and pattern matching, it should be possible to identify residues in natural organic matter (Simpson al. 2004a). Also the development of HR-MAS technique opens new perspectives to characterize molecular species at the solid-aqueous interface, under conditions closer to those of natural environment. Another challenge is now to investigate the major question of bio-availability of organic pollutants, mainly for bacteria. In this context we are currently using $^{1}H$ HR-MAS to monitor directly in situ the degradation of benzothiazole-2-sulfonic acid by *Rhodococcus* strains in the presence of anionic clays (Haroune 2003).

## References

Achtnich C, Fernandes E, Bollag JM, Knackmuss HJ, Lenke H (1999) Covalent binding of reduced metabolites of [$^{15}N_3$]TNT to soil organic matter during bioremediation process analysed by $^{15}N$ NMR spectroscopy. Environ Sci Technol 33:4448-4456

Ahmad R, Kookana RS, Alston AM, Skjemstad JO (2001) The nature of soil organic matter affects sorption of pesticides. 1. Relationships with carbon chemistry as determined by $^{13}C$ CPMAS NMR spectroscopy. Environ Sci Technol 35:878-884

Benoit P, Preston C (2000) Transformation and binding of $^{13}C$- and $^{14}C$-labelled atrazine in relation with straw decomposition in soil. Eur J Soil Sci 51:43-54

Berns A, Vinken R, Bertmer M, Breitschwerdt A, Schäffer A (2005) Use of $^{15}N$-depleted artificial compost in bound residue studies. Chemosphere (in press)

Bruns-Nagel D, Knicker H, Drzyzga O, Butehorn U, Steinbach K, Gemsa D, Von Low E (2000) Characterization of $^{15}N$-TNT residues after an anaerobic/aerobic treatment of soil/molasses mixtures by solid-state $^{15}N$ NMR spectroscopy. 2. Systematic investigation of whole soil and different humic fractions. Environ Sci Technol 34:1549-1556

Combourieu B, Haroune N, Besse P, Sancelme M, Delort AM (2003) $^{1H}$ nuclear magnetic resonance : a tool to study biodegradative pathways of organic pollutants in *Mycobacterium* and *Rhodococcus* isolates. Res. Adv. Microbiol 3:1-22

Combourieu B, Inacio J, Delort AM, Forano C (2001) Differentiation of mobile and immobile pesticides on anionic clays by $^{1}H$ HR MAS NMR spectroscopy. Chem Commun 21:2214-2215

Cook RL, McIntyre DD, Langford CH, Vogel HJ (2003) A comprehensive liquid-state heteronuclear and multidimensional NMR study of Laurentian fulvic acid. Environ Sci Technol 37:3935-3944

Delort AM, Combourieu B (2000) Microbial degradation of xenobiotics. In: Barbotin JN, Portais JC (eds) NMR in Microbiology: theory and applications, Chapter 19. Horizon Scientific, UK, pp 411-430

Delort AM, Combourieu B (2001) *In situ* $^{1}H$-NMR study of the biodegradation of xenobiotics: application to heterocyclic compounds. J Industr Microbiol Biotechnol 26:2-8

Delort AM, Combourieu B, Haroune N (2004) Nuclear Magnetic Resonance studies of interactions between organic pollutants and soil components: a review. Environ Chem Lett 1:209-213

Eriksson J, Frankki S, Shchukarev A, Skyllberg U (2004) Binding of 2,4,6-trinitrotoluene, aniline and nitrobenzene to dissolved and particulate soil organic matter. Environ Sci Technol 38:3074-3080

Grivet JP, Delort AM, Portais JC (2003) NMR and microbiology: from physiology to metabolomics. Biochimie 85:823-840

Gunasekara AS, Simpson MJ, Xing B (2003) Identification and characterization of sorption domains in soil organic matter using structurally modified humic acids. Environ Sci Technol 37:852-8

Haider K, Spiteller M, Wais A, Fild M (1993) Evaluation of the binding mechanism of anilazine and its metabolites in soil organic matter. Int J Environ Anal Chem 53:125-137.

Haroune N (2003) Métabolisme de benzothiazoles par des souches de *Rhodococcus:* étude par RMN in situ. University Blaise Pascal, France

Käcker T, Haupt ET, Garms C, Francke W, Steinhart H (2002) Structural characterization of humic acid-bound PAH residues in soil by $^{13}C$-CPMAS-NMR-spectroscopy: Evidence of covalent bonds. Chemosphere 48:117-131

Kang S, Xing B (2005) Phenanthrene sorption to sequential extracted soil humic acids and humans. Environ Sci Technol 39:134-140

Keeler C, Maciel GE (2003) Quantitation in the solid-state $^{13}C$ NMR analysis of soil and organic soil fractions. Anal Chem 75:2421-2432

Khalaf M, Kohl SD, Klumpp E, Rice JA, Tombacz E (2003) Comparison of sorption domains in molecular weight fractions of a soil humic acid using solid-state $^{19}F$ NMR. Environ Sci Technol 37:2855-2860

Khol SD, Toscano PJ, Wenhua H, Rice JA (2000) Solid-state $^{19}F$ NMR investigation of hexafluorobenzene sorption to soil organic matter. Environ Sci Technol 34:204-210

Kile DE, Wershaw RL, Chiou CT (1999) Correlation of soil and sediment organic matter polarity to aqueous sorption of nonionic compounds. Environ Sci Technol 33:2053-2056

Knicker H, Bruns-Nagel D, Drzyzga O, Von Low E, Steinbach K (1999) Characterization of $^{15}N$-TNT residues after an anaerobic/aerobic treatment of soil/molasses mixtures by solid-state $^{15}N$ NMR spectroscopy. 1. Determination and optimization of relevant NMR spectroscopic parameters. Environ Sci Technol 33:343-349

Knicker H, Achtnich C, Lenke H (2001) Solid-state nitrogen-15 nuclear magnetic resonance analysis of biologically reduced 2,4,6-trinitrotoluene in a soil slurry remediation. J Environ Quality 30:403-410

Knicker H, Hatcher PG, Gonzalez-Vila FJ (2002) Formation of heteroaromatic nitrogen after prolonged humification of vascular plant remains as revealed by nuclear magnetic resonance spectroscopy. J Environ Quality 31:444-449

Knicker H (2003) Incorporation of $^{15}N$-TNT transformation products into humifying plant organic matter as revealed by one- and two-dimensional solid state NMR spectroscopy. Sci Total Environ 308:211-220

Kögel-Knabner I (2000) Analytical approaches for characterizing soil organic matter. Org Geochem 31:609-625

Kulikova NA, Perminova IV (2002) Binding of atrazine to humic substances from soil, peat and coal related to their structure. Environ Sci Technol 36:3720-4

Nanny MA, Minear RA, Leenheer JA (1998) Nuclear magnetic resonance in environmental chemistry. Oxford University Press, New-York

Piotto M, Bourdonneau M, Furrer J, Bianco A, Raya J, Elbayed K (2001) Destruction of magnetization during TOCSY experiments performed under magic angle spinning: effect of radial B1 inhomogeneities. J Magn Reson 149:114-118

Preston C (2001) Carbon-13 solid-state NMR of soil organic matter-using the technique effectively. Can J Soil Sci 81:255-270

Salloum MJ, Chefetz B, Hatcher PG (2002) Phenanthrene sorption by aliphatic-rich natural organic matter. Environ Sci Technol 36:1953-1958

Shih YH, Wu SC (2002) Sorption kinetics of selected volatile organic compounds in humin. Environ Sci Technol 21:2067-2074

Simpson AJ (2001) Multidimensional solution state NMR of humic substances: a practical guide and review. Soil Sci 166:795-809

Simpson AJ, Kingery WL, Shaw DR, Spraul M, Humpfer E, Dvortsak P (2001) The application of $^{1}H$ HR-MAS NMR spectroscopy for the study of structures and associations of organic components at the solid-aqueous interface of a whole soil. Environ Sci Technol 35:3321-3325

Simpson AJ, Salloum MJ, Kingery WL, Hatcher PG (2002) Improvements in the two-dimensional nuclear magnetic resonance spectroscopy of humic substances. Environ Quality 31:388-392

Simpson MJ, Chefetz B, Hatcher PG (2003a) Phenanthrene sorption to structurally modified humic acids. J Environ Quality 32:1750-1758

Simpson AJ, Kingery WL, Hatcher PG (2003b) The identification of plant derived structures in humic materials using three-dimensional NMR spectroscopy. Environ Sci Technol 37:337-342

Simpson AJ, Lefebvre B, Moser A, Williams A, Larin N, Kvasha M, Kingery, WL, Kelleher B (2004a) Identifying residues in natural organic matter through spectral prediction and pattern matching of 2D NMR datasets. Magn Reson Chem 42:14-21

Simpson MP, Simpson AJ, Hatcher P (2004b) Noncovalent interactions between aromatic compounds and dissolved humic acids examined by nuclear magnetic resonance spectroscopy. Environ Toxicol Chem 23:355-362

Tao T, Maciel GE (2002) Interaction of methyl bromide with soil. Environ Sci Technol 36:603-607

Thiele S, Fernandes E, Bollag JM (2002) Enzymatic transformation and binding of labeled 2,4,6-trinitrotoluene to humic substances during an anaerobic/aerobic incubation. J Environ Quality 31:437-444

Thorn KA, Kennedy KR (2002) $^{15}$N NMR investigation of the covalent binding of reduced TNT amines to soil humic acid, model compounds, and lignocellulose. Environ Sci Technol 36:3787-3796

Thorn KA, Pennington JC, Hayes CA (2002) $^{15}$N NMR investigation of the reduction and binding of TNT in an aerobic bench scale reactor simulating windrow composting. Environ Sci Technol 36:797-805

Witte EG, Philipp H, Vereecken H (1998) Binding of $^{13}$C-labelled 2-aminobenzothiazoles to humic acid as derived from $^{13}$C NMR spectroscopy. Org Chem 29:1829-1835

Witte EG, Philipp H, Vereecken H (2002) Study of enzyme-catalysed and non-catalysed interactions between soil humic acid and $^{13}$C-labelled 2-aminobenzothiazoles using solid-state $^{13}$C NMR spectroscopy. Org Chem 33:1727-1735

Xing B (2001) Sorption of naphthalene and phenanthrene by soil humic acids. Environ Pollution 111:303-309

Zhu D, Hyun S, Pignatello JJ, Lee L (2004) Evidence of $\pi$–$\pi$ interactions between $\pi$donor aromatic compounds and $\pi$acceptor sites in soil organic matter through pH effects on sorption. Environ Sci Technol 38:4361-4368

# 2 Modelling

# Incorporating geomicrobial processes in reactive transport models of subsurface environments

P. Regnier, A.W. Dale, C. Pallud, Y. van Lith, S. Bonneville, C. Hyacinthe, M. Thullner, A.M. Laverman, P. Van Cappellen

Department of Earth Sciences - Geochemistry – P.O. Box 80021 – 3508 TA Utrecht – The Netherlands – Tel: +31-30-2535005 – Fax: +31-30-2535302 – e-mail: p.regnier@geo.uu.nl

## Abstract

Reactive-transport models aim at a comprehensive, quantitative and, ultimately, predictive treatment of biogeochemical transformations and mass transfers in the subsurface. Not only do they provide environmental simulation tools, they can also be used to test new theoretical concepts or hypotheses. A major goal of the geochemistry group in Utrecht is to incorporate complex, microbially-driven reaction networks in reactive transport models, through a close collaboration between modelers and experimentalists. This paper gives an overview of some of the research activities we are carrying out in this area.

## Keywords

Biogeochemical reaction networks; Michaelis-Menten kinetics; microbial growth; thermodynamic constraints; sensitivity

## Introduction

The chemical structure and evolution of subsurface environments are largely determined by the activity and the diversity of the resident microbial populations (Chapelle 2000). Microbial metabolic reactions modify the acid-base and electrolytic properties of pore waters as well as the composition of particulate plus colloidal matter in soils, sediments and aquifers. Comprehensive modeling of reactive transport in subsurface environments therefore requires mathematical expressions that predict the rates at which microorganisms consume and produce chemical constituents.

Microbial processes typically exhibit saturation behavior with respect to substrates that the organisms extract from their surroundings. That is, with increasing concentration of a limiting substrate, the rate ultimately reaches a maximum value. This behavior is captured by the so-called Michaelis-Menten (or Monod) rate expression, which is commonly used to represent microbial reaction pathways in biogeochemical reactive transport models (e.g., Mayer et al. 2001). The Michaelis-Menten model, however, has mainly been verified for pure cultures of bacteria consuming soluble substrates. We present the results of recent experimental studies, which illustrate that the Michaelis-Menten formulation also holds for natural microbial communities and when solid-phase substrates are utilized.

In subsurface environments, several groups of microorganisms often compete for the same substrate(s). A classic example of competition is the degradation of soil or sediment organic matter by heterotrophic microorganisms that couple organic carbon oxidation to the reduction of alternative terminal electron acceptors (TEAs). The principal TEAs are generally used sequentially in the order $O_2$, $NO_3^-$, Mn(IV), Fe(III), and $SO_4^{2-}$, leading to a characteristic redox zonation. Here, we analyze the competition for organic energy substrates by microorganisms using different TEAs, within the framework of a reaction network model based on Michaelis-Menten kinetics.

There is growing interest in understanding the functioning of microbial communities in environments with a low supply of energy substrates, e.g., deep aquifers or deep sediments. In such areas, microorganisms function close to their bioenergetic limit, meaning that thermodynamic constraints must be included in the metabolic rate description. A simple extension of the Michaelis-Menten rate expression is employed that ensures thermodynamic consistency. The extended rate model is applied to anaerobic oxidation of methane (AOM) by sulfate, a biogeochemical process carried out by symbiotic associations of methanogenic archaea and sulfate-reducing bacteria (e.g., Hoehler et al. 1994).

As knowledge about microbial processes and interactions expands, increasingly detailed representations of geomicrobial reaction systems are being developed. This creates the need for robust approaches for parameter identification and calibration. Here, we use a factorial analysis to identify the parameters that most significantly affect the output of our AOM reaction network model. This type of global sensitivity analysis should, ideally, be incorporated into the design of experimental studies of complex geomicrobial systems. It may also assist in the derivation of simplified rate equations of geomicrobial reactions which, in turn, can be included in reactive transport models that simulate geomicrobial activity in subsurface environments.

## Geomicrobial kinetics

The Michaelis-Menten rate expression was originally derived for enzyme-catalyzed reactions in which one reactant species, the substrate, is transformed into a product via an enzyme-substrate complex. It relates the initial reaction rate to the total enzyme concentration (assumed constant), and the initial substrate concentration. The rate of product formation reaches a maximum ($R_{max}$) when all enzyme molecules are complexed with the substrate. The substrate concentration at which the rate equals half of $R_{max}$ is referred to as the affinity constant ($K_m$), which is a measure of the affinity of the enzyme for the substrate: the smaller $K_m$, the stronger the affinity. In its simplest form, the Michaelis-Menten rate expression is written as

$$R = R_{max} \cdot \frac{[S]}{K_m + [S]} \tag{1}$$

where $R$ is the rate of the enzyme-catalyzed reaction and $[S]$ the substrate concentration.

The Michaelis-Menten rate expression is frequently applied to multistep reactions carried out by complex microbial communities (e.g., Martin-Nieto et al. 1992, Ulrich et al. 2003). In Fig. 1, for example, rates of bacterial sulfate reduction in an intertidal, freshwater sediment from the Scheldt river (Appels, Belgium) are plotted as a function of the dissolved sulfate concentration. The rates were measured in continuous flow-through reactors containing intact slices of sediment (Roychoudhury et al. 1998), and with dissolved sulfate being supplied to the resident microbial community via the inflow solution. The electron donors used by the sulfate reducing bacteria, however, are generated inside the reactor through hydrolysis and fermentation of sedimentary organic matter. Despite the fact that different groups of microorganisms are involved, as well as a variety of intermediate chemical species, Fig. 1 shows that the Michaelis-Menten expression provides a good description of the net consumption of the TEA.

In a number of environmentally-relevant cases, microorganisms rely on solids as direct substrates. For example, in the absence of appropriate electron shuttles, the dissimilatory reduction of ferric oxyhydroxides requires direct contact between the mineral and the bacteria. Hence, the Fe(III) reduction kinetics are expected to depend on mineral properties that affect cell adhesion, such as the shape, specific surface area and surface charge of the mineral particles (e.g., Roden et al. 1996), as well as on aggregation of the cells or production of extracellular polymers. Confronted with so many potentially rate-controlling variables, it is reasonable to question whether the simple formalism of Eq. 1 is still adequate.

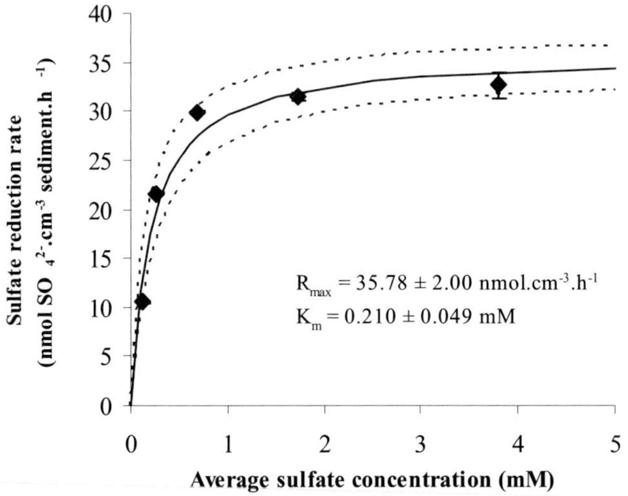


**Fig. 1.** Steady-state sulfate reduction rates in the 2-4 cm depth interval of an intertidal freshwater sediment (Appels, Belgium). Rates are measured in a continuous flow through reactor containing undisturbed sediment slices, as a function of the average sulfate concentration in the reactor. The solid line is the best fit of the Michaelis-Menten expression (Eq. 1) to the experimental data.

We have systematically investigated the dependence of the rate of reduction of Fe(III) mineral phases by the iron reducing bacterium, *Shewanella putrefaciens* (Bonneville et al. 2004). The minerals investigated so far include two different hematites, lepidocrocite, 6-lines ferrihydrite, amorphous Fe(III) (hydr)oxide and ferric phosphate. For all the solids, the dependence of the Fe(III) reduction rate on the concentration of the Fe(III) substrate fits the Michaelis-Menten equation, as illustrated in Fig. 2a for 6-lines ferrihydrite.

The same experimental approach was used to determine rates of microbial Fe(III) reduction by *S. putrefaciens* of natural, iron rich sediment from the Scheldt river. Also in this case, the rate exhibits saturation behavior with respect to the amount of reactive Fe(III) present in the sediments (Fig. 2b). The maximum Fe(III) reduction rate *per* cell is of the same order of magnitude as those obtained for ferrihydrite (Fig. 2a) and amorphous

Fe(III) (hydr)oxide, indicating that in the sediment, the ferric mineral(s) are highly reactive and can be used by the resident iron reducing microorganisms.

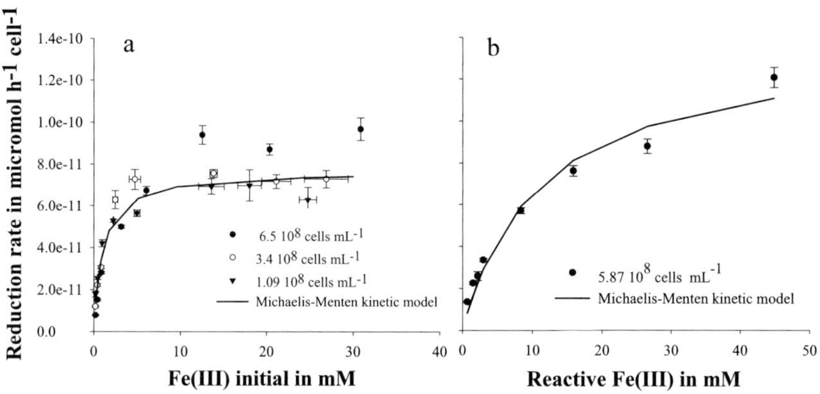


**Fig. 2. a.** Rate of reductive dissolution of 6-lines ferrihydrite by the facultative anaerobic bacterium *Shewanella putrefaciens*, as a function of the initial concentration of Fe(III), at three different bacterial cell densities. The rates plotted are normalized *per* cell (specific rates). The experiments were carried out under strict anaerobic conditions at pH 7, with excess lactate as electron donor. The solid line corresponds to the best fit of the Michaelis-Menten expression to the experimental data. **b.** Same as in a., but now with a natural iron-rich, intertidal freshwater sediment (Appels, Belgium) as Fe(III) source. The rate scale is identical to that in a. The rates are plotted versus the concentration of reactive Fe(III), defined as the fraction of total iron extractable by ascorbate at circumneutral pH (Hyacinthe and Van Cappellen 2004).

The above studies, and others, confirm that the Michaelis-Menten rate expression provides a general description of the utilization of external substrates by microorganisms. Hence, it forms the logical starting point for the development of mathematical representations of geomicrobial processes in reactive transport models. In order to account for changes in the microbial community structure, microbial groups, e.g., sulfate or iron reducing bacteria, can be included explicitly in Eq. 1, by expanding $R_{max}$ as $R_{max} = v_{max}\ B$, where $B$ is the biomass of a particular microbial group and $v_{max}$ the maximum specific reaction rate *per* unit biomass or cell. The standard approach consists in solving a conservation equation for $B$ that expresses biomass growth and decay. This implies an additional level of model parameterization, such as specifying the maximum specific growth rate, growth yield and decay constant of the microorganisms (section 3).

In contrast to natural porous media, it is, in principle, straightforward to manipulate and monitor the cell density (biomass) in pure culture experiments (e.g., Fig. 2a). For natural samples, traditional culture-enrichment techniques introduce biases because not all microorganisms grow equally well in the culture media. Recently-developed molecular probing techniques are rapidly increasing our insight in the diversity and community structure of natural microbial populations (e.g., Amman et al. 1995). However, relatively few studies have so far directly linked the abundance of specific functionalities or microbial groups to geomicrobial reaction rates in subsurface environments.

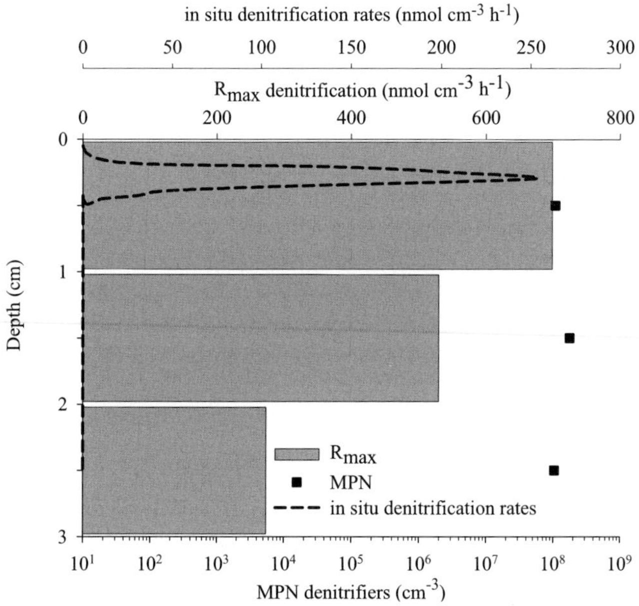


**Fig. 3.** Depth distributions of $R_{max}$ for nitrate reduction obtained using plug-flow through reactor experiments, abundance of denitrifiers (MPN) and *in situ* nitrate reduction rates in an intertidal freshwater sediment (Appels, Belgium). *In situ* nitrate reduction rates were derived from time-series $N_2O$ microprofiles in sediment cores treated with the $N_2O$ reductase inhibitor acetylene.

In a recent study of nitrate reduction in estuarine sediments (Laverman et al. submitted), no correlation was observed between the abundance of nitrate reducing bacteria, obtained with the Most Probable Number (MPN) method, and the maximum rate, $R_{max}$, of $NO_3^-$ reduction (Fig. 3). Furthermore, *in situ* denitrification was restricted to the top 5 mm of the sediment core, as $NO_3^-$ did not penetrate deeper in the sediment. Potential denitrification activity, however, extended much deeper (Fig. 3). Thus, the total population density was not a reliable predictor of the *in situ* or potential denitrification rates. A major challenge ahead will be to determine which measurable characteristics of natural microbial populations directly relate to potential reaction rates. Until then, caution must be exerted when interpreting biomass terms appearing in the rate expressions of geomicrobial reactions. At best, $B$ should be considered an effective, or apparent, biomass. The uncertainties surrounding *in situ* biomasses are reminiscent of those encountered in water-rock reactive transport problems, when reactive mineral surface areas must be estimated from bulk mineral abundances.

One drawback of Eq. 1 is that it treats the transformation of the substrate as an irreversible process, whose rate only vanishes when the concentration of the substrate reaches zero. However, energy-generating (catabolic) microbial processes, for instance dissimilatory sulfate and iron(III) reduction, only proceed when the energy yield exceeds some metabolic threshold. An extension of the Michaelis-Menten expression for a multi-substrate catabolic process, which accounts for bioenergetic limitation and possible inhibition, is

$$R = v_{\max} \cdot B \cdot \left( \prod_i \frac{[S_i]}{K_{\mathrm{m}}^i + [S_i]} \right) \cdot \left( \prod_j \frac{K_{\mathrm{in}}^j}{K_{\mathrm{in}}^j + [IN_j]} \right) \cdot \left( 1 - \exp\left( \frac{\Delta G^*}{\chi \mathrm{RT}} \right) \right) \tag{2}$$

where $K_{\mathrm{m}}^i$ is the affinity (or half-saturation) constant of the $i$-th substrate $S_i$, $K_{\mathrm{in}}^j$ is the inhibition constant associated with the $j$-th inhibitor $IN_j$, $\Delta G^*$ is the Gibbs energy of reaction, corrected for the threshold energy, $\chi$ is the average stoichiometric number of the reaction (Jin and Bethke 2002), and R and T are the gas constant and absolute temperature.

Alternative formulations of the extended Michaelis-Menten expression have been proposed. Nonetheless, they all convey the same essential features of microbial redox transformations. That is, the rate, $R$, of a given redox pathway depends on the abundance and maximum activity levels of the microorganisms involved, the availability and affinity of the cells for the direct electron donor and acceptor species, the presence of inhibitors that may inactivate the pathway, and the Gibbs energy yield of the corresponding redox transformation. The latter is harvested by the microorganisms for their growth and maintenance. Note that in Eq. 2, accumulation of reaction products limits the reaction rate *via* the effect on the Gibbs energy of reaction.

Inclusion of the last term on the right-hand-side (RHS) of Eq. 2 is particularly important when dealing with geomicrobial systems operating close to their thermodynamic limit (Jin and Bethke 2002). An example is anaerobic oxidation of methane (AOM) by sulfate (Hoehler et al. 1994). The catabolic reactions carried out by the methane-oxidizing archaea (MOA) and sulfate-reducing bacteria (SRB) are

$$\text{MOA}: \ CH_4 + 3H_2O \rightarrow 4H_2 + HCO_3^- + H^+ \tag{3a}$$

$$\text{SRB}: \ 4H_2 + SO_4^{2-} + H^+ \rightarrow HS^- + 4H_2O \tag{3b}$$

$$\text{Net}: \ CH_4 + SO_4^{2-} \rightarrow HS^- + H_2O + HCO_3^- \tag{3c}$$

where $H_2$ represents the interspecies electron shuttle. This reaction is energetically not very favorable; its standard Gibbs energy (298 K, 1 atm) is only –16.9 kJ *per* mole $CH_4$. Hence, the rate of AOM is expected to be sensitive to fairly small changes in saturation state.

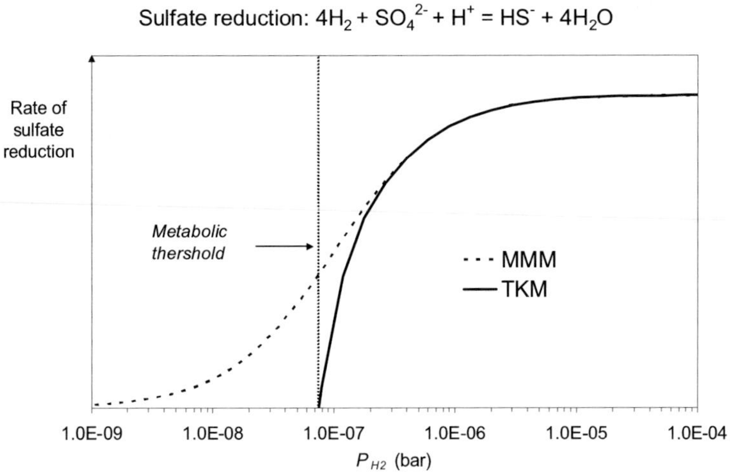


**Fig. 4.** Rate of sulfate reduction coupled to $H_2$ oxidation (Eq. 3b) during AOM, as a function of the partial pressure of $H_2$. MMM refers to the standard Michaelis-Menten model (Eq. 1); TKM is the combined Thermodynamic-Kinetic Model, which explicitly relates metabolic rates to the Gibbs energy of reaction (Eq. 2). Far from equilibrium, when $H_2$ is abundant, the two models converge.

Following Jin and Bethke (2002), we have developed a mixed kinetic-bioenergetic model of AOM where Eq. 2 is used to compute the rates of the various microbial pathways involved (section 3). The simulations show that the higher kinetic efficiency of SRB causes this reaction (Eq. 3b) to proceed close to equilibrium, while methane oxidation (Eq. 3a) remains relatively far from equilibrium. The latter is a consequence of the low partial pressure of $H_2$, because of the efficient utilization of $H_2$ by SRB. The dependence of the rate of sulfate reduction on the Gibbs energy of reaction (i.e., the last term on the RHS of Eq. 2) also prevents the concentration of $H_2$ from dropping below a minimum value, which corresponds to the threshold Gibbs energy requirement of the SRB (Fig. 4).

## Biogeochemical reaction networks

Microbial reactions in soils, sediments and aquifers are part of extended reaction networks, which include biotic and abiotic transformation processes. The substrates and products of these biogeochemical pathways are highly coupled, leading to a complex dynamics which is best captured by the means of models. The first step in developing a reaction network model is pathway identification. For each biogeochemical pathway there is a set of reactants and products, and, for biotic pathways, also a biomass, *B*. Once all the pathways have been identified, the complete set of chemical species and biomasses is uniquely defined.

A schematic representation of a typical biogeochemical reaction network is given in Fig. 5. It includes some of the major reaction pathways encountered in anaerobic subsurface environments, where degradation of organic matter sustains a metabolically versatile microbial community capable of using a variety of TEAs. A number of different modeling approaches have been proposed to simulate the sequential utilization of TEAs during the breakdown of organic matter (see, e.g., Curtis 2003). Using a batch reaction model based on the extended Michaelis-Menten approach, the competition between respiratory pathways in a reaction network such as the one in Fig. 5 is shown to depend mainly on the values of the affinity constants for the electron donors and, to a lesser degree, on the growth yields of the various microbial populations.

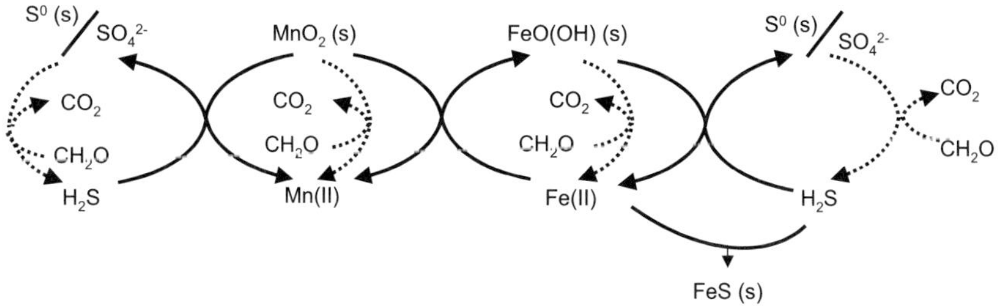


**Fig. 5.** Simplified C-Mn-Fe-S reaction network in anaerobic environments with organic carbon as the energy source. The dashed lines correspond to the organic carbon respiration pathways with Mn(IV), Fe(III) and sulfate as TEAs.

The model results further show that when labile organic matter is abundant the affinity constants for the TEAs also play an important role in the dynamics of the reaction network shown in Fig. 5. In particular, these affinity constants influence the relative distribution of TEA utilization over competing reduction pathways. For example, a decrease in the affinity constant of Fe(III) (hydr)oxides for iron reducing bacteria may shift Fe(III) reduction to reaction with sulfide, which consequently promotes organic matter degradation by SRB. In general, with increasing availability of labile organic matter, there is also growing spatial and temporal overlap among respiratory pathways using different TEAs, as can be demonstrated experimentally (Fig. 6).

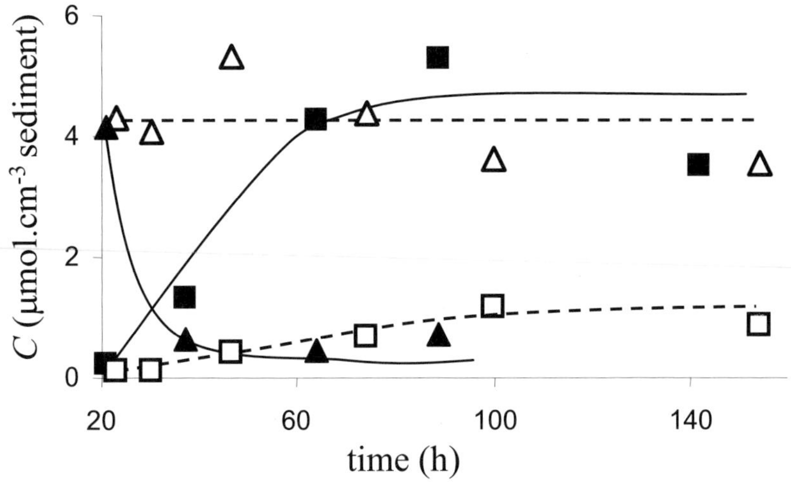


**Fig. 6.** Measured concentrations of sulfate (triangles) and $Fe^{2+}$ (squares) in the aqueous phase of an unamended sediment slurry incubation (open symbols and dashed lines) compared to a slurry incubation amended with lactate (filled symbols and solid lines). The temporal overlap among iron and sulfate reduction increases when the labile organic substrate is added (amended slurry). The sediment used is from an intertidal site of the Scheldt river (Appels, Belgium).

One of the key capabilities of models is to help determine which parameters have the highest impact on the performance of geomicrobial reaction systems. If the model contains only a few adjustable parameters, then a brute force sensitivity analysis, where each parameter is varied successively, while holding the others constant, may still be an option. However, even when relatively simple geomicrobial reaction networks are considered, the number of model parameters rapidly makes this approach obsolete. We illustrate this for the case of AOM introduced in section 2.

AOM occurs in marine sediments where upward diffusing methane meets downward diffusing sulfate. The microbial consortia of MOA and SRB carrying out AOM depend on a close syntrophic association and energy sharing for their survival (Eq. 3). In particular, the SRB are essential for the removal of $H_2$ produced by the MOA. To simulate the growth of these microorganisms and the turnover of methane in marine sediments, we have developed a reaction network including not only methane oxidation (Eq. 3a) and sulfate reduction coupled to $H_2$ oxidation (Eq. 3b), but also methanogenesis, acetogenesis, fermentation, and sulfate reduction coupled to oxidation of organic fermentation products (Dale et al. submitted). Equation 2 is used to compute the rates of the individual pathways, while the biomasses of the various microbial groups, $B_i$, are calculated by

$$\frac{dB_i}{dt} = \sum_j Y_{ij} \cdot R_j - \mu_e \cdot B_i \tag{4}$$

where the sum is taken over all the pathways used by the $i$-th microbial group to generate energy, $Y_{ij}$ stands for the growth yield of the microbial group by the $j$-th catabolic pathway, $R_j$ is the rate of the $j$-th pathway in units of biomass obtained with Eq. 2, and $\mu_e$ is the specific endogenous decay rate of the microorganisms.

Our current reaction network model embodies the still fragmentary conceptual understanding of how AOM communities function. Yet the model already requires over 40 independent parameters, which reflects the inherently complex nature of reaction systems in which several microbial groups participate. Clearly, there is a need to identify those parameters, or combination of parameters, that most significantly affect the model output. These are also the parameters that deserve particular attention when planning future laboratory and field experiments. Factorial analysis has proven to be a simple, but powerful, approach in this respect, and applications are commonplace in environmental engineering (Montgomery 1996).

A baseline simulation is first conducted, using the best available estimates of the parameters. Subsequently, the parameter values are perturbed and the deviation of some defined model response from the baseline simulation is monitored. The sensitivity range for each parameter is chosen based on experimental data or informed guesswork. An example of results of a factorial analysis of the AOM reaction network is shown on a normal probability diagram in Fig. 7. Each point on the diagram represents the effect of a parameter on the steady-state rate of AOM, which is used as model response. Points lying on the straight line indicate a normally distributed model response. Therefore, the effect of the corresponding parameter on the model output is indistinguishable from normally distributed (random) error. Points lying away from the line identify the significant parameters at the sensitivity levels tested.

Using a coastal marine sediment scenario undergoing transient seasonal forcings, the most sensitive parameters for AOM are found to be those belonging to sulfate-reduction coupled to $H_2$ oxidation and methane production from acetate (Fig. 7). In addition to the affinity constants of $H_2$ and acetate for the SRB and methanogens, respectively, and the maximum growth rates of these two groups of microorganisms, the model response is also sensitive to the thermodynamic state descriptors,

$$F_T = \left(1 - \exp\left(\frac{\Delta G^*}{\chi \mathrm{RT}}\right)\right) \quad (5)$$

of the two pathways. The latter observation reinforces the importance of including thermodynamic constraints in the kinetic descriptions of the metabolic pathways in the reaction network.

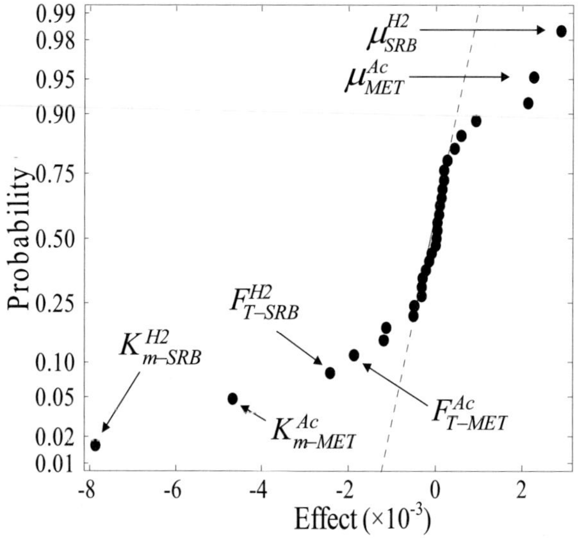


**Fig. 7.** Normal probability plot of a factorial analysis of AOM: the X-axis corresponds to the estimated effects of the various parameters, the Y-axis to the probability. Parameters exhibiting the greatest departures from the straight line most significantly affect the model output. These parameters are the specific maximum growth rates (μ) of sulfate reducing bacteria (SRB) and methanogens (MET), and the affinity constants ($K_m$) and thermodynamic descriptors ($F_T$, Eq. 5) of the pathways of sulfate reduction coupled to $H_2$ and methanogenesis from acetate (Ac).

The sensitivity analysis further shows that the overall rate of AOM in this transient scenario is not sensitive to the kinetic and thermodynamic parameters related to the methane oxidation pathway (Eq. 3a) and MOA biomass. That is, the explicit representation of these parameters, and the related variables, does not improve the performance of the reaction network model. This result, which at first may seem counterintuitive, would not have been obtained without considering the entire set of microbial pathways and their interactions. The factorial analysis therefore yields important information for the derivation and parameterization of simplified mathematical reaction models that can be incorporated in diagenetic reactive transport codes.

## Geomicrobiology in reactive-transport models

The simulation of reactive-transport in earth systems requires the incorporation of geomicrobial reaction networks into a transport framework. In one dimension, this coupling is represented by the conservation equation:

$$\frac{\partial c_i}{\partial t} = D\frac{\partial^2 c_i}{\partial x^2} - v\frac{\partial c_i}{\partial x} + R_i \tag{6}$$

where $c_i$ is the concentration of chemical species $i$ in the reaction network at time $t$ and position $x$, $D$ is an effective diffusion coefficient used to represent all diffusive/dispersive transport processes, and $v$ is the velocity associated with advective transport. $R_i$ represents the sum of all reactive processes affecting the concentration of species $i$. The explicit representation of the microbial community requires similar conservation equations for the biomasses, $B_i$, where $R_i$ accounts for the growth and decay of the microorganisms (Eq. 4).

As an illustration of reactive-transport applications, two environments with contrasting transport intensities and rates of geomicrobial processes are presented. In the first example, the process of AOM in passive continental shelf sediments is simulated. The mixed kinetic-bioenergetic model for $CH_4$, $SO_4^{2-}$, acetate and $H_2$ is coupled to transport by substituting the rate expressions discussed earlier (Eq. 2) into the mass conservation equations (Eq. 6). Figure 8 shows computed vertical profiles of $CH_4$, $SO_4^{2-}$ and $H_2$, values of the thermodynamic ($F_T$, Eq. 5) and kinetic ($F_K$) limitations of reaction 3a alongside rates of AOM. The term $F_K$ corresponds to the first term in parenthesis on the RHS of Eq. 2, with $CH_4$ as the substrate. Figure 8 illustrates that AOM only occurs within a well-defined zone of synergistic kinetic and thermodynamic drive. Above the AOM zone, kinetic inhibition ($F_K$ = 0) arises from a decrease in $CH_4$ concentrations. Below the zone, thermodynamic inhibition ($F_T$ = 0) arises from the build-up of end products ($H_2$) due to reduced $H_2$ consumption by SRB (Hoehler et al. 1994).

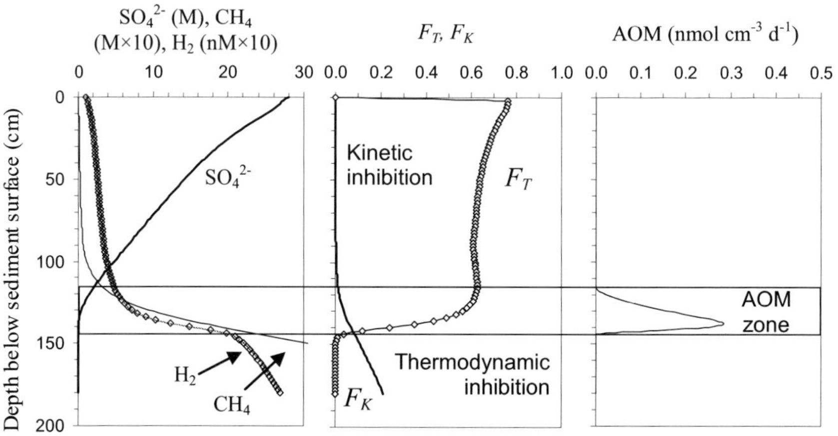


**Fig. 8. a.** Depth-concentration profiles of $SO_4^{2-}$, $CH_4$ and $H_2$ in a marine sediment, **b.** Values for the thermodynamic ($F_T$) and kinetic ($F_K$) limitation terms for methane oxidation (Eq. 3a), and **c.** Rate of AOM. $F_T$ and $F_K$ vary between 0 (total growth limitation) and 1 (no growth limitation).

The second example is based on experimental data on the degradation of dissolved organic carbon (DOC) in a sand column experiment flushed with artificial groundwater (von Gunten and Zobrist 1993). The reaction network involves a set of biotic and abiotic geochemical reactions similar to the one illustrated in Fig. 5. The rates of the different microbial degradation pathways of DOC are described by Eq. 2, assuming $F_T$ = 1. Results show that both the total carbon degradation rate and the relative contribution of each redox metabolic pathway vary in space and time (Fig. 9). In this advection-dominated flow system, the redox rates correlate with the transient microbial biomass distributions. In contrast, and in agreement with experimental findings (Fig. 3), model simulations of bioturbated aquatic sediments reveal no correlation between bacterial abundances and rates of corresponding redox processes (Thullner et al. in press). These examples highlight the importance of the transport regime in controlling the relationship between microbial community structure and microbial activity in subsurface environments.

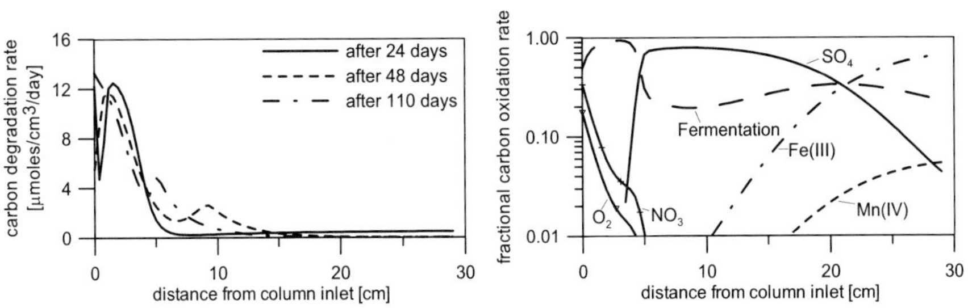


**Fig. 9.** Simulated carbon degradation rates in a sand column experiment fed with groundwater containing lactate as labile carbon source. Left: total carbon degradation rate along the column at different times; Right: relative contribution of different organic carbon degradation pathways (as indicated by the TEA used) to the total degradation rate after 110 days.

## Conclusions

The Michaelis-Menten equation has proven remarkably successful in reproducing the rates of complex geomicrobial processes. It therefore provides a robust basis for the quantitative modeling of biogeochemical reaction networks in subsurface environments. Extensions of the basic Michaelis-Menten equation allow one to take into account the effects of multiple substrates, microbial growth, inhibitors and bioenergetic constraints. The development of detailed geomicrobial reaction network models helps to close the gap between fundamental research in geochemistry, microbial ecology and molecular biology, and application-oriented reactive transport modeling.

## Acknowledgments

This research was financially supported by several grants from the Netherlands Organization for Scientific Research (NWO) and by the European Commission (METROL Contract no. EVK3-CT-2002-00080).

## References

Amman RI, Ludwig W, Schleifer KH (1995) Phylogenetic identification and in situ detection of individual microbial cells without cultivation. Microbiological Reviews, Vol. 59,pp 143-169

Bonneville S, Van Cappellen P, Behrends T (2004) Microbial reduction of iron(III) oxyhydroxides: Effects of mineral solubility and availability. Chemical Geology, Vol. 212, pp 255-268

Chapelle FH (2000) Ground-Water Microbiology and Geochemistry. 2nd Ed. John Wiley & Sons, N.Y.

Curtis GP (2003) Comparison of approaches for simulating reactive solute transport involving organic degradation reactions by multiple terminal electron acceptors. Computers & Geosciences, Vol.29, pp 319-329

Dale AW, Regnier P, Van Cappellen P (2005) Bioenergetic controls of anaerobic oxidation of methane in organic-rich marine sediments: A sensitivity analysis. American Journal of Science. Submitted

Hoehler TM, Alperin MJ, Albert DB, Martens CS (1994) Field and laboratory studies of methane oxidation in an anoxic marine sediment - evidence for a methanogen-sulfate reducer consortium. Global Biogeochemical Cycles, Vol. 8, pp 451-463

Hyacinthe C, Van Cappellen P (2004) An authigenic iron phosphate phase in estuarine sediments: composition, formation and chemical reactivity. Marine Chemistry, Vol.91, pp 227-251

Jin Q, Bethke CM. (2002) Kinetics of electron transfer through the respiratory chain. Biophysical Journal, Vol.83, pp 1797-1808

Laverman AM, Van Cappellen P, Meile C, Wieringa EB (2005) Rate distribution of denitrification in an estuarine sediment: Microprofiling and reactive transport modeling. Limnology & Oceanography. Submitted

Martin-Nieto J, Flores E, Herrero A (1992) Biphasic kinetic behavior of nitrate reductase from heterocystous, nitrogen-fixing cyanobacteria. Plant Physiology, Vol.100, pp 157-163

Mayer KU, Benner SG, Frind EO, Thornton SF, Lerner DN (2001) Reactive transport modeling of processes controlling the distribution and natural attenuation of phenolic compounds in a deep sandstone aquifer. Journal of Contaminant Hydrology, Vol.53, pp 342-368

Montgomery DC (1996) Design and Analysis of Experiments, Wiley, N.Y.

Roden EE, Zachara JM (1996) Microbial reduction of crystalline iron(III) oxides: Influence of oxide surface area and potential for cell growth. Environmental Science and Technology, Vol.30, pp 1618-1628

Roychoudhury AN, Viollier E, Van Cappellen P (1998) A plug flow-through reactor for studying biogeochemical reactions in undisturbed aquatic sediments. Applied Geochemistry, Vol. 13, pp 269-280

Thullner M, Van Cappellen P, Regnier P (2005) Modeling the impact of microbial activity on redox dynamics in porous media. Geochimica Cosmochimica Acta, In press

Ulrich GA, Breit GN, Cozzarelli IM, Suflita JM (2003) Sources of sulfate supporting anaerobic metabolism in a contaminated aquifer. Environmental Science and Technology, Vol. 37, pp 1093-1099

von Gunten U, Zobrist J (1993) Biogeochemical changes in groundwater-infiltrations systems: column studies. Geochimica et Cosmochimica Acta, Vol. 57, pp 3895-3906

# Consequences of Different Kinetic Approaches for Simulation of Microbial Degradation on Contaminant Plume Development

D. Schäfer[1], A. Manconi[2], S. Grandel[1], A. Dahmke[1]

[1] Dept. of Applied Geology, Christian-Albrechts-Universität zu Kiel, Germany
[2] Dept. of Land Engineering, Università degli Studi di Cagliari, Italy

## Abstract

This study investigates the consequences of assuming different kinetic approaches for calculation of microbial degradation on plume development in a simple case of natural attenuation on field scale. If the required parameters are properly chosen, all approaches can simulate similar plumes for a particular given time step. The differences of contaminant concentrations in the plumes are small and would not attract attention in a natural aquifer. On long term prognoses the kinetics result in very different plumes: A complex Monod approach considering microbial growth prognoses a further spreading of the plume, compared to a first order rate law, which results in a short and early stationary plume. Other approaches show plumes between these two extremes. On the other hand, the forecasts for plumes assuming Monod kinetics are similar, even if different values for parameterization are chosen. The reason for this insensibility is, that degradation is not limited by microbial kinetics in the simulation, but by dispersive mixing. Simplifying approaches may have few and well determinable parameters, but they are not suited for proper prognoses if they neglect the prerequisite, that contaminant and electron acceptor have to be present for a reaction.

## Keywords

bioremediation; modeling; microbial kinetics; prognosis; reliability

## Introduction

For natural attenuation (NA) of organic pollutants in aquifers microbial degradation has to be considered as the most important destructive process (Wiedemeier et al. 1999). Biological degradation is based on a complex system of enzymatic reactions in cell organelles and transfer processes through membranes, but most of these processes are not yet completely understood (Alexander 1994, Chapelle 1993). Hence, various simplifying mathematical approaches exist to simulate microbial degradation (Koch 1998, Schmidt et al. 1985, Simkins and Alexander 1984).

Monod-kinetics is one of the most complex approaches, it considers changes in microbial population density, maximum uptake rates and non-linear dependency of turnover on substrate concentrations. This approach is quite sophisticated, but it is rarely used to simulate natural attenuation (Schäfer 2001), because it is numerically extensive and because it requires a large number of hardly determinable parameters. Michaelis-Menten kinetics (Borden and Bedient 1986) is a simplification of Monod-kinetics, since it neglects changes of microbial population and therefore spatial and temporal variation of maximum turnover rate. It is annoying, that these two approaches are often not distinguished in literature, e.g. Borden and Bedient (1986) investigate "Monod-kinetics", but assume a constant concentration of microorganisms, i.e. in fact it is a Michelis-Menten kinetics. In order to reduce numerical effort and parameter demand other simplifying approaches are used, like first-order (McNab Jr. and Narasimhan 1994) and zero-order rate laws (Nielsen and Christensen 1994) or kinetics are completely neglected by assuming instantaneous chemical equilibrium (Borden and Bedient 1986).

The simplification of complex kinetics yields the risk of erroneous results and misinterpretations. From direct comparison of mathematical functions it is known, that first-order as well as zero-order rate laws can overestimate the degradation rate up to several thousand percent compared to Michaelis-Menten kinetics (Bekins et al. 1998). This overestimation has been confirmed in 1D biofilm reactor models (Bonomo et al. 2000). Simkins and Alexander (1984) investigate the application range of 6 different kinetic approaches for interpretation of measured benzoate degradation in batch experiments. These and other investigations (Kelly et al. 1996, Schmidt et al. 1985) were performed in batch or column studies, where short timescales are chosen intending to observe microbial kinetics. Additionally in batch experiments electron acceptors required for contaminant degradation are often available in surplus and do never limit degradation. These conditions are not transferable to natural attenuation on site scale, where long timescales are investigated and where the presence of electron donor as well as electron acceptor can inhibit degradation.

On larger scales (tank experiment ~1m) Michaelis-Menten kinetics have found to be more accurate than first order degradation to simulating aerobic BTX degradation (Jean et al. 2002). On site scale instantaneous chemical equilibrium has shown to be a good approximation of Michaelis-Menten kinetics (Rifai and Bedient 1990) or Monod kinetics (Koussis et al. 2003) for "fast" reactions or "long" time scales (i.e. high Damköhler number), but it causes erroneous results, if the reaction is slower than expected. Instantaneous equilibrium is also a better approach to simulating intrinsic remediation in field applications than a first order decay model (Newell et al. 1995).

These individual comparisons on different scales demonstrate, that the approach chosen to calculate microbial degradation will influence the simulation of plume development and possibly the decision, if NA is applicable. This study will compare all the different approaches discussed above and apply them to a simple scenario of aerobic toluene degradation in a homogeneous aquifer with stationary flow. These simplified conditions make sure, that all concentration changes are attributed to kinetic effects and not to heterogeneity or complex boundary conditions, allowing a direct comparison of different approaches. The spreading of the contaminant plume is simulated assuming the diverse kinetic approaches and differences of the generated plumes and in forecast plume evolution are discussed. The following questions will be attributed: How severe are the differences of simulated plumes on field scale, if diverse kinetics or chemical equilibrium are assumed? Is it possible to identify the underlying kinetics from measured concentrations? Which consequences has the assumption of a kinetic approach on the prognosis of future plume development? How different are ill parameterized complex Monod kinetics compared to a well parameterized approach with simplified kinetics?

## Methods

The numerical model TBC (Schäfer et al. 1998) is able to simulate all the approaches discussed above. This ensures, that flow- and transport conditions are exactly the same for all simulations.

The simulated aquifer with dimensions of 120m x 40m x 4m (XYZ) is discretized in cells of 1m x 1m x 1m. The homogeneous conductivity is 2 $m \cdot d^{-1}$ in horizontal and 0.2 $m \cdot d^{-1}$ in vertical direction. A gradient of 2.5‰ in X-direction and an effective porosity of 33% result in a flow velocity of

1.5 cm·d$^{-1}$. In the beginning of the simulation the aquifer is aerobic with an oxygen concentration of 0.3 mmol·l$^{-1}$, which is also found in the inflow boundary. An immobile toluene NAPL phase of 3m x 10m is assumed on the groundwater table, dissolving into the mobile pore water via a gradient law

$$\partial C_{mobile} / \partial t = k \cdot \left( C_{equilibrium} - C_{mobile} \right) \tag{1}$$

with a maximum solubility $C_{equilibrium}$ of 5.6 mmol·l$^{-1}$ and a constant dissolution velocity $k$ of 0.001 d$^{-1}$. For transport of toluene and oxygen a longitudinal dispersivity $\alpha_L$ of 0.90 m, transversal dispersivities in horizontal and vertical direction of $\alpha_{Th}$=0.09 m and $\alpha_{Tv}$=0.009 m, respectively, and a molecular diffusion of 7.4E-5 m²·d$^{-1}$ are assumed.

The first numerical approach to describe degradation considers microbial growth via Monod-kinetics. It is used to calculate concentrations after 2000, 4000 and 6000 days (~ 16 y) simulation time. The other approaches require kinetic parameters, which can not be derived from the Monod-parameters. Hence, these unknown parameters and degradation rates are chosen such that they reproduce the concentration distribution after 2000 d. Main focus was set to obtain the same plume lengths. With the parameterized kinetics the future behavior of the plume is calculated for 4000d and 6000d simulation time. The prognostic capabilities of the approaches are evaluated by comparing the simulated plumes.

For application of Monod-kinetics a group of immobile toluene oxidizing aerobic microbes is ubiquitary present in low concentrations (5E-4 mmolC·kg$^{-1}$). Growth of this microbial group $X$ follows a "double Monod-kinetics" (Bae and Rittmann 1996) with two multiplied Monod-terms for toluene and oxygen, allowing a microbial growth and degradation only if toluene and oxygen are present. A constant rate simulates biomass decay e.g. due to endogenous respiration. Net growth rate is growth minus decay, resulting in a permanent consumption of substrates even if biomass is constant.

$$\frac{\partial X}{\partial t} = v_{max} \cdot X \cdot \frac{C_{toluene}}{MC_{toluene} + C_{toluene}} \cdot \frac{C_{O_2}}{MC_{O_2} + C_{O_2}} - v_{dec} \cdot X \tag{2}$$

The values for maximum growth velocity $v_{max}$, half velocity concentrations $MC$ for toluene and oxygen, and decay rate $v_{dec}$ assumed for this reaction are summarized in table 1. The consumption of toluene is coupled to microbial growth by a Yield-coefficient $Y_{toluene}$.

$$\frac{\partial C_{toluene}}{\partial t} = -\frac{1}{Y_{toluene}} \cdot v_{max} \cdot X \cdot \frac{C_{toluene}}{MC_{toluene} + C_{toluene}} \cdot \frac{C_{O_2}}{MC_{O_2} + C_{O_2}} \tag{3}$$

An analogous formulation is used to calculate oxygen consumption, the Yield-coefficient is calculated from the stoichiometric oxygen demand.

$$C_6H_5CH_3 + 9\ O_2 + 3\ H_2O \Rightarrow 7\ H_2CO_3 \tag{4}$$

Assuming a constant microbial activity over time and space and the presence of a sufficiently large amount of oxygen, Eq. 3 can be transformed to:

$$\frac{\partial C_{toluene}}{\partial t} = -k_M \cdot \frac{C_{toluene}}{MC_{toluene} + C_{toluene}} \tag{5}$$

This equation is a Michaelis-Menten kinetics with respect to toluene concentration. The main difference is the assumption of a spatially and temporally constant microbial population. The half-velocity concentration for toluene $MC_{toluene}$ is taken from table 1, the lumped maximum degradation rate $k_M$ is chosen such that the length of the toluene plume after 2000d simulation time is equal to the previous plume calculated with Monod-kinetics.

For small toluene concentrations the denominator of the Monod-term is approximately $MC_{toluene}$ and Eq. 5 can be transformed to a first order rate law:

$$\frac{\partial C_{toluene}}{\partial t} = -k_1 \cdot C_{toluene} \tag{6}$$

For high toluene concentrations the Monod-term is nearly one and Eq. 5 can be substituted by a zero order rate law:

$$\frac{\partial C_{toluene}}{\partial t} = -k_0 \tag{7}$$

If the reaction is fast compared to the transport velocity (high Damköhler number), equilibrium is achieved and kinetics of the reaction can be neglected. A law of mass action is formulated according to the stoichiometric equation for toluene oxidation (Eq. 4):

$$\frac{C_{H_2CO_3}{}^{7}}{C_{toluene} \cdot C_{O_2}{}^{9}} = K \tag{8}$$

Equilibrium is far on product side, causing a reaction of toluene and oxygen until one reaction partner is nearly completely consumed. The approach requires no fitting of kinetic parameters.

In order to test the relevance of accurate parameterization of Monod-kinetics, two parameter sets deviating from the values in table 1 are chosen. The first set of parameters generally overestimates microbial degradation by a factor of 2 in all parameters, while the second scenario underestimates degradation (table 2).

The simulated plumes are displayed and analyzed with Surfer®. In addition to simple optical comparison the following more objective criteria are used to compare the simulated plumes.

1. Contaminant mass in the plume at simulated time steps (measure for contaminant degradation)
2. Plume volume with contaminant concentrations >0.02 mmol·l-1 (measure for total plume volume)
3. Plume volume with contaminant concentrations >0.40 mmol·l-1 (measure for high contaminated areas)
4. Plume length

Contaminant mass in the plume is calculated by the simulation program TBC. Surfer® determines the areas of high concentrations (>0.40 $mmol \cdot l^{-1}$) and total plume area (>0.02 $mmol \cdot l^{-1}$) in every simulated aquifer layer. Multiplication with the layer thicknesses results in plume volumes. The limiting values for "high" and "low" concentrations were chosen arbitrarily. Plume length is defined as the distance between downstream boundary of the NAPL source zone and the forefront of the plume defined by the concentration isoline of 0.02 $mmol \cdot l^{-1}$.

**Table 1.** Parameters of Monod-kinetics

| | | | | | |
|---|---|---|---|---|---|
| $v_{max}$ | 1.0 $d^{-1}$ | $MC_{toluene}$ | 1E-5 $mmol \cdot l^{-1}$ | $Y_{toluene}$ | 0.100 $mol \cdot mol^{-1}$ |
| $v_{dec}$ | 0.1 $d^{-1}$ | $MC_{O2}$ | 1E-5 $mmol \cdot l^{-1}$ | $Y_{O2}$ | 0.086 $mol \cdot mol^{-1}$ |

**Table 2.** Monod-parameters for over- or underestimation of microbial degradation

| overestimation of microbial degradation | | | | | |
|---|---|---|---|---|---|
| $v_{max}$ | 2.0 $d^{-1}$ | $MC_{toluene}$ | 0.5E-5 $mmol \cdot l^{-1}$ | $Y_{toluene}$ | 0.200 $mol \cdot mol^{-1}$ |
| $v_{dec}$ | 0.1 $d^{-1}$ | $MC_{O2}$ | 0.5E-5 $mmol \cdot l^{-1}$ | $Y_{O2}$ | 0.172 $mol \cdot mol^{-1}$ |
| underestimation of microbial degradation | | | | | |
| $v_{max}$ | 0.5 $d^{-1}$ | $MC_{toluene}$ | 2E-5 $mmol \cdot l^{-1}$ | $Y_{toluene}$ | 0.050 $mol \cdot mol^{-1}$ |
| $v_{dec}$ | 0.1 $d^{-1}$ | $MC_{O2}$ | 2E-5 $mmol \cdot l^{-1}$ | $Y_{O2}$ | 0.043 $mol \cdot mol^{-1}$ |

## Results and discussion

If Monod kinetics is considered, the highest bacteria population and therefore main degradation potential evolves at the upstream boundary of the NAPL contamination, where toluene and oxygen are mixed advectively (no figure). Downstream of the source degradation is only found in the dispersive mixing zone of the plume fringe. Outside of the plume no degradation occurs due to a lack of toluene, inside of the plume oxygen is depleted.

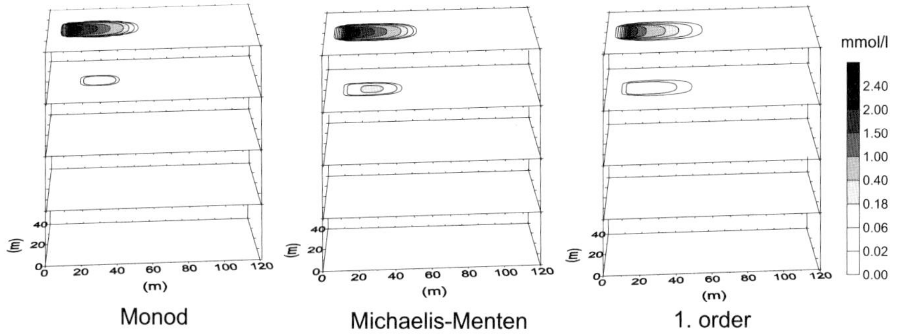


**Fig. 1.** Toluene concentrations in the simulated 5 aquifer layers after 2000d of simulation time for three different approaches for microbial degradation

Only toluene concentrations are considered for comparison of different simulation approaches. The unknown parameters of the approaches are chosen such that the toluene plumes after 2000d are similar to the plume calculated with the Monod approach. No simplifying approach was able to reproduce this plume exactly in all details (Fig. 1). Hence, we decided to choose the parameters such that the plume lengths after 2000d are the same. This means, that other criteria like contaminant masses of the plumes (see Fig. 4) are different already at t=2000d. Figure 1 shows the toluene concentrations after 2000d for Monod-kinetics (Eq. 3), Michaelis-Menten kinetics (Eq. 5), and first order degradation (Eq. 6). Difference plots allow a more detailed comparison of the simulated toluene concentrations after 2000d (Fig. 2). Positive values indicate, that the appropriate approach overestimates contaminant concentrations compared to the Monod approach.

Michaelis-Menten kinetics satisfactorily reproduce the plume simulated with Monod-kinetics. Upstream of the contaminant source toluene concentrations are overestimated with respect to the Monod approach (up to +8.2% $C_{max}$), due to neglecting a high bacteria population and increased uptake rates. The same is found for first order degradation, though in predominant part of the plume concentrations are underestimated (up to -37.3% $C_{max}$). Zero order degradation generally overestimates concentrations (up to +9.4% $C_{max}$) in the whole plume regarding the Monod approach. Toluene concentrations calculated assuming chemical equilibrium agree well with the Monod kinetics, only in the plume front concentrations are underestimated (up to -1.0% $C_{max}$) due to neglecting the lag-time of bacteria.

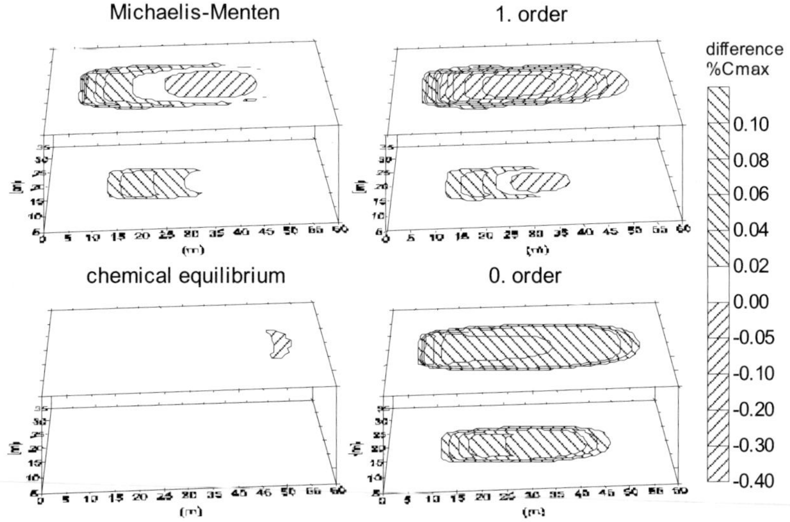


**Fig. 2.** Difference plots toluene concentrations "appropriate approach – Monod-approach", detail of the upper two aquifer layers after 2000d simulation time

While plume shapes agree quite well after a simulation time of 2000d, the prognosis of the future plume development shows increasing deviations. Figure 3 shows the toluene concentrations after 6000d for the approaches discussed above. The plume simulated assuming a Michaelis-Menten kinetic is shorter and not as wide (horizontally and vertically) as the plume simulated for Monod-kinetics. The first order degradation approach overestimates degradation in high contaminated areas resulting in a short and already stationary plume.

The prognoses of all kinetic approaches are evaluated by calculated plume lengths, contaminant masses and plume volumes with toluene concentrations >0.40 $mmol \cdot l^{-1}$ or >0.02 $mmol \cdot l^{-1}$ (Fig. 4). Compared to Monod-kinetics all other approaches generally underestimate the plume length. The plume simulated with first order degradation reaches steady state after 4000d. The other approaches show increasing plume lengths, but slower than under consideration of microbial growth.

The mass of toluene in the plume is differs already after 2000d of simulation time (Fig. 3), because the distributions of contaminant concentrations in the plumes are different for most approaches. With spreading of the plume contaminant mass increases, but to a different degree for each approach. Due to varying toluene distributions the volumes of the plume containing toluene concentrations >0.40 mmol·l$^{-1}$ or >0.02 mmol·l$^{-1}$ vary and develop differently over time (Fig. 4).

The chosen over- and underestimation of microbial degradation assuming Monod-kinetics has only negligible consequences on plume spreading. Plume lengths, masses and volumes are nearly identical to the calculation with the original Monod parameters (Fig. 4).

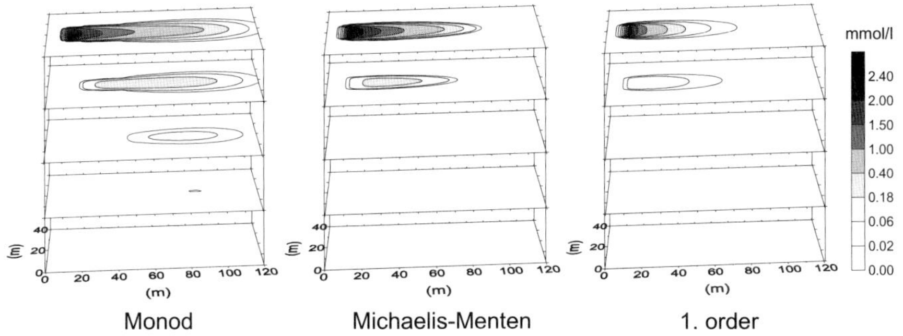


**Fig. 3.** Prognosed toluene concentrations in the simulated 5 aquifer layers after 6000d simulation time for three approaches for microbial degradation

## Conclusions

As expected the numerical approaches for calculation of microbial degradation result in similar, but not identical plumes after 2000d simulation time. Due to sparse observational networks, measuring errors, mixing in full screened observation wells or in heterogeneous flow fields it is unlikely that concentration differences in the range of ±20% can be detected. Hence, most of the kinetics investigated can sufficiently reproduce a single snapshot of measured contaminant concentrations and it seems to be impossible to identify the underlying kinetics from one concentration distribution only. The only exception is first order degradation with significant concentration differences compared to the other approaches. Since this approach is quite often used, either degradation in real applications

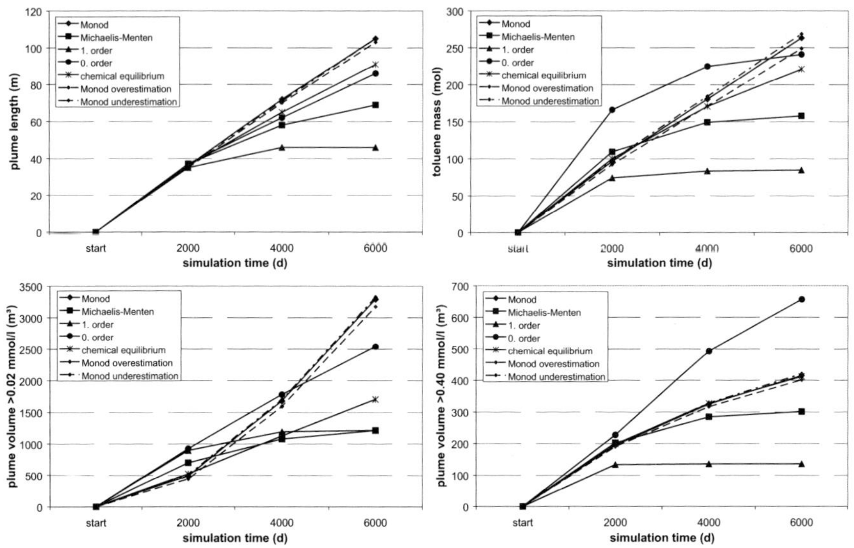


**Fig. 4.** Plume development over time for the assumed numerical approaches

follows first order kinetics (in contrast to our simulation) or even these larger concentration differences are not detectable in real sites.

With increasing time the simulated plumes behave different, hence measured concentration time series may increase the chance to detect the underlying kinetics. If the assumed kinetic is not appropriate, it may be able to reproduce the plume at a specific time, but the long term prognosis will fail. All investigated simplifying approaches overestimate degradation compared to Monod kinetics and forecast shorter or early stationary plumes, what is an alarming result with regard to risk assessment.

The main argument against the more complex Monod approach is the large number of unknown parameters, which are often not well-definable. The systematic over- and underestimation of microbial degradation by changing all parameter values by a factor of 2 has no relevant consequences on contaminant plume behavior, that is the simulation results are not very sensitive to these parameter changes. The plumes calculated with ill parameterized Monod-kinetics agree better with the original plume than the approaches with simplifying kinetics.

The reason for this insensibility is, that in the simulation aerobic degradation is not limited by microbial kinetics, but by the dispersive mixing of aerobic water in the aquifer and toluene contaminated water inside of the plume, what is a well known phenomenon for dissolved contaminants and electron acceptors (Cirpka et al. 1999, Klenk and Grathwohl 2002). Monod kinetics take care, that microbial degradation occurs only if toluene and oxygen are present at the same time and at the same place. The same applies to the assumption of chemical equilibrium. Hence, Monod kinetics and chemical equilibrium show very similar plumes as expected from Koussis et al. (2003), Borden and Bedient (1986) or Rifai and Bedient (1990), though the concentration distribution inside of the plumes is not exactly the same (Fig. 3).

Simplifying kinetic approaches do not consider, that contaminant and electron acceptor have to be present. A first order rate law, for example, calculates degradation like radioactive decay, depending on contaminant concentrations only, and simulates degradation even in the centre of the plume, where no oxygen is present (Jean et al. 2002; Newell et al. 1995). But the same applies for simple Michaelis-Menten kinetics or zero order degradation. All these approaches overestimate degradation, causing non-conservative prognoses, because the rate limiting step (dispersion) is neglected. Simplifying approaches can only be successful, if they represent the kinetics of the degradation limiting step, e.g. if microbial consumption of Fe(III)-minerals follows a (pseudo) zero order rate law, the whole process of Fe(III)-reduction and contaminant oxidation should be zero order.

The presented results are based on the simple case of one non-sorbing contaminant and a single electron acceptor. For more complex (i.e. most natural) systems the assumption of a single degradation rate or the equilibrium approach are hardly applicable. If several electron acceptors are present, degradation rates will change depending on the redox reaction. In chemical equilibrium exclusively the thermodynamically most favorable oxidant is consumed, what does not correspond to the observation, that more than one degradation reaction can occur at the same time (Lovley and Phillips 1987, Lyngkilde and Christensen 1992). Microbial degradation can be inhibited by the presence of toxic substances, what is not considered in simplifying approaches, and some thermodynamically favorable reactions simply do not occur, if the catalyzing microorganisms are not present. Kinetic effects also have to be considered, if competitive reactions occur, e.g. if oxygen is consumed by aerobic bacteria and pyrite oxidation or if organic contaminants are adsorbed or degraded simultaneously.

Hence, simplifying approaches should only be assumed for prognoses, if the assumptions required to drive them are in fact fulfilled. E.g. if electron acceptors are never limiting degradation, at small contaminant concentrations and sufficient amounts of active microorganisms catalyzing the reaction, the assumption of a first order rate law may be successful. If degradation is limited by slow mixing of electron acceptor and –donor, the simulation approach has to make sure that a reaction only occurs, if both reactants are present – what is a trivial but often neglected demand. In aquifers with one electron acceptor, one donor and high Damköhler numbers chemical equilibrium can be an adequate simplifying approach. For more complex systems Monod kinetics are recommended. Though we cannot conclude from the simulations, that Monod-kinetics simulate the “real” degradation processes and plume spreading, it definitely considers many of the processes known from extensive laboratory studies. The problem, that the Monod parameters cannot be exactly identified, has no significant consequences on the simulated plume at least on the simple system investigated.

## Acknowledgments

This work was partially funded by the German Ministry of Science and Education (BMBF) as part of the “Virtual Aquifer” project (02WN0389). The stay of the coauthor A. Manconi at the University Kiel was funded by the EU student exchange program ERASMUS.

## References

Alexander M (1994) Biodegradation and Bioremediation. Academic Press Inc., San Diego

Bae W, Rittmann BE (1996) A structured model of dual-limitation kinetics. Biotechnology Bioengineering 49:683-689

Bekins BA, Warren E, Godsy EM (1998) A Comparison of Zero-Order, First-Order, and Monod Biotransformation Models. Ground Water 36:261-268

Bonomo L, Pastorelli G, Quinto E (2000) Simplified and Monod kinetics in one-dimensional biofilm reactor modelling: a comparison. Water Science & Technology 43:295-302

Borden RC, Bedient PB (1986) Transport of Dissolved Hydrocarbons Influenced by Oxygen-Limited Biodegradation, 1. Theoretical Development. Water Resources Research 22:1973-1982

Chapelle FH (1993) Ground-Water Microbiology and Geochemistry. John Wiley & Sons Inc., New York

Cirpka OA, Frind EO, Helmig R (1999) Numerical simulation of biodegradation controlled by transverse mixing. J Cont Hydr 40:159-182

Jean JS, Tsai CL, Ju SH, Tsao CW, Wang SM (2002) Biodegradation and transport of benzene, toluene, and xylenes in a simulated aquifer: comparison of modelled and experimental results. Hydrological Processes 16:3151-3168

Kelly WR, Hornberger GM, Herman JS, Mills AL (1996) Kinetics of BTX biodegradation and mineralization in batch and column systems. J Cont Hydr 23:113-132

Klenk ID, Grathwohl P (2002) Transverse vertical dispersion in groundwater and the capillary fringe. J Cont Hydr 58:11-128

Koch AL (1998) The Monod model and its alternatives. In: Koch AL, Robinson JA, Milliken GA (ed) Mathematical modeling in microbial ecology. Chapman & Hall, New York, pp 62-93

Koussis AD, Pesmajoglou S, Syriopoulou D (2003) Modelling biodegradation of hydrocarbons in aquifers: when is the use of the instantaneous reaction approximation justified? J Cont Hydr 60:287-305

Lovley DR, Phillips EJP (1987) Competitive mechanisms for inhibition of sulfate reduction and methane production in the zone of ferric iron reduction in sediments. Applied Environmental Microbiology 11:2636-2641

Lyngkilde J, Christensen TH (1992) Redox zones of a landfill leachate pollution plume (Vejen, Denmark). J Cont Hydr 10:273-289

McNab Jr. WW, Narasimhan TN (1994) Modeling reactive transport of organic compounds in groundwater using a partial redox disequilibrium approach. Water Resources Research 30:2619-2635

Newell CJ, Winters JA, Miller RN, Gonzales J, Rifai HS, Wiedemeier TH (1995) Modeling Intrinsic Remediation With Multiple Electron Acceptors: Results From Seven Sites. Proceedings: Petroleum Hydrocarbons and Organic Chemicals in Ground Water. Houston, Texas. www.gsi-net.com/Publications/meamodel.pdf

Nielsen PH, Christensen TH (1994) Variability of biological degradation of aromatic hydrocarbons in an aerobic aquifer determined by laboratory batch experiments. J Cont Hydr 15:305-320

Rifai HS, Bedient PB (1990) Comparison of Biodegradation Kinetics With an Instantaneous Reaction Model for Groundwater. Water Resources Research 26:637-645

Schäfer D, Schäfer W, Kinzelbach W (1998) Simulation of reactive processes related to biodegradation in aquifers: 1. Structure of the three-dimensional reactive transport model. J Cont Hydr 31:167-186

Schäfer W (2001) Predicting natural attenuation of xylene in groundwater using a numerical model. J. Cont. Hydr. 52:57-83

Schmidt SK, Simkins S, Alexander M (1985) Models for the Kinetics of Biodegradation of Organic Compounds Not supporting Growth. Applied and Environmental Microbiology 50:323-331

Simkins S, Alexander M (1984) Models for Mineralization Kinetics with the Variables of Substrate Concentration and Population Density. Applied and Environmental Microbiology 47:1299-1306

Wiedemeier TH, Rifai HS, Newell CJ, Wilson JT (1999) Natural Attenuation of Fuels and Chlorinated Solvents in the Subsurface. John Wiley & Sons, Inc., New York

# Natural Attenuation in the unsaturated zone and shallow groundwater: coupled modeling of vapor phase diffusion, biogeochemical processes and transport across the capillary fringe

U. Maier, P. Grathwohl

Center for Applied Geoscience, University of Tübingen

Corresponding author: Sigwartstr. 10 – 72076 Tübingen – Germany – Tel.: +49.70712975429 – fax: +49.70715059 – e-mail: grathwohl@uni-tuebingen.de

## Abstract

The unsaturated zone including the capillary fringe is a very dynamic and active environment for biogeochemical processes. Modeling of fate and transport of organic pollutants has to account for vapor phase and sewage water transport coupled to the biogeochemical processes occurring. This study presents results from numerical simulations validated with data of a well-controlled field experiment on volatilization of a multi-component organic mixture in the unsaturated zone. Sensitivity analyses show that the overall biodegradation rates depend mainly on properties of the organic pollutants such as Henry's Law constant, the soil water content, and on the individual degradation rate constants or temperature. Low Henry's law constants result in relatively high biodegradation rates whereas compounds with high vapor pressure und low water solubility are lost to the atmosphere. The contaminant transfer rates into groundwater are relatively small, but lead locally to concentrations above the legal limit in the capillary fringe region. Contaminant transport by seepage water is just minor compared to diffusive vapor phase fluxes of volatile compounds.

## Keywords

Unsaturated zone; organic contaminants; groundwater pollution risk; scenario-specific modelling

## Introduction

Contaminated land poses a serious problem with respect to soil quality and the risk of spreading of pollutants into other compartments of the environment. A major concern at most contaminated sites is the risk of groundwater pollution by organic and inorganic compounds. Since the remediation of all of the contaminated sites is economically not feasible in many countries, groundwater risk assessment procedures are needed for the ranking of sites, decision making on further use and remedial actions. Of special interest is Monitored Natural Attenuation (Wiedemeyer 1999), which so far, however, is mostly applied in polluted aquifers and only to a very limited extent in the unsaturated zone.

At sites where petroleum products are handled or stored, contamination of the unsaturated soil zone is frequently found. Hydrocarbons can reach the groundwater by transport with percolating water and by spreading in the soil-gas. Degradation processes can limit the spreading in the unsaturated soil zone and - in the best case - restrict the contamination to the unsaturated zone. The objective of this study was to use numerical experiments to elucidate the processes/parameters which are relevant in contaminant spreading and thus for groundwater risk assessment. The results are compared to data from a well controlled field test performed at the Værløse Airforce Base, Denmark (Christophersen et al. 2005). Sensitivity analyses were performed accounting for physical-chemical properties of volatile fuel constituents and properties of the soil.

## Model and base scenario

The numerical model MIN3P (Mayer 1999) uses the finite volume method and allows for the calculation of vapour phase transport and unsaturated flow in the vadose zone. Groundwater transport processes and mass transfer across the capillary fringe can be simulated as well and a variable number of geochemical reactions such as biodegradation processes can be handled (Mayer et al. 2002). Density driven gas advection was not included in the simulations. It is often considered a significant transport process for compounds with very high vapour pressures in highly permeable materials such as coarse gravel. Hughes-Conant et al. (1996), however, showed that in sandy material density induced advection is of moderate importance only at high vapour densities.

The base scenario for the model was adapted to a well controlled field test in Denmark, where a small volume (approx. 10.2 l) of a multi-component NAPL (non-aqueous phase liquid) was buried and monitored over a time period of one year (Christophersen et al. 2005). Site characteristics with respect to grain size distribution, geochemistry, permeability, etc. were implemented in the model as precisely as possible. The site has a vadose zone of 3.5 meters thickness, consisting of sandy and not too heterogeneous material. The source was emplaced in the subsurface at 1 m depth. The synthetic hydrocarbon mixture consisted of 13 volatile to semi-volatile kerosene compounds and one tracer (freon). Volatilization of the mixture and vapour phase diffusion of single compounds in disturbed samples was investigated independently by Wang et al. (2003). Biodegradation processes were analysed by Höhener et al. (2002) and used as starting values for model calibration . Monitoring of the contaminant spreading in soil air and groundwater samples in a one-year measurement campaign allowed a comparison to modelled data. A compilation of model input parameters is given in Tab. 1 and a vertical cross section of the model domain in Fig. 1. To reduce computational time, several sensitivity analyses were performed in 1D and 2D, whereas the model was fitted to field data in 3D.

**Table 1.** Compilation of important parameters of the field site model.

| | |
|---|---|
| Porosity | 34 - 40 % |
| Van Genuchten parameters | $\theta_{res}$ = 2.5 – 3.6 %, $\alpha$ = 2.9 – 3.2 $m^{-1}$, $n$ = 1.84 – 2.97 (porosity) |
| Climatic conditions | Temperature & precipitation daily means from weather station at the site, evapotranspiration according to Haude |
| Hydraulic properties (aquifer) | Flow velocity $v$: 0.2 – 0.3 m $day^{-1}$, hydraulic conductivity $K$: 1.2 10-4 m/s |
| Diffusion coefficients $D_{aq}$ and $D_{air}$ | 8.10-10 $m^2$ $s^{-1}$ (aqueous) and 7.3.10-6 $m^2$ $s^{-1}$ (gaseous phase) |
| NAPL composition | 10.2 l of kerosene with 13 compounds & 1 tracer (freon CF113) |

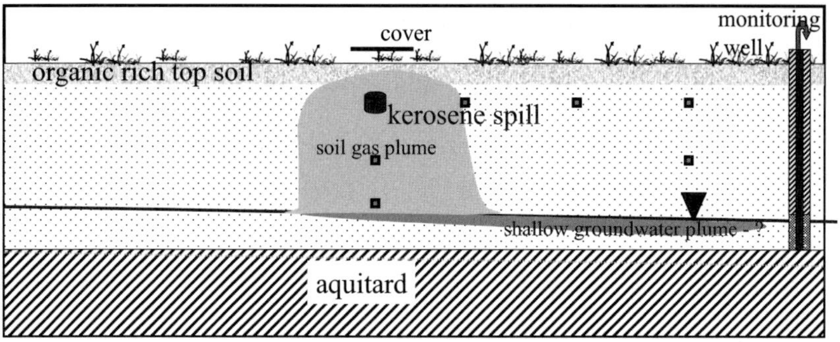


**Fig. 1.** Conceptual model, base case used for sensitivity analysis in numerical simulations.

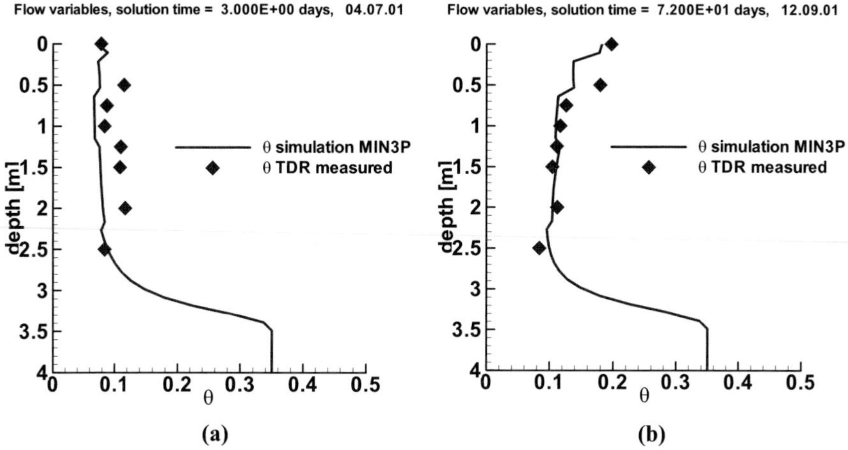


**Fig. 2.** Soil moisture profile at day 3 and at day 72 of the experiment, simulated using MIN3P and experimentally determined from TDR measurements. $\theta$: volumetric soil water content.

Measured soil temperature profiles were implemented in the model as well as precipitation and evapotranspiration at the site, as transient boundary conditions with daily averaged values. That way, infiltration and soil water contents were calculated by MIN3P using the Van-Genuchten approach for variably saturated flow. Potential evapotranspiration ETPot was calculated according to Haude (Hölting 1992). Transient behaviour of soil water content calculated that way was compared to TDR (time-domain-reflectometry) measurements close to the site (Fig. 2). Profile (a) was found in a dry period in the summer (start of the experiment), whereas (b) was observed during a period of rainfall (autumn). TDR signals, however, were found to overestimate soil water contents somewhat compared to exact gravimetric determination.

## Results - sensitivity analyses

To evaluate the importance of compound properties, soil and meteorological parameters for the contaminant behaviour, sensitivity analyses on biodegradation rate constants, Henry's law constant H, aqueous saturation concentration CSat for the different compounds, temperature, soil water content and possible anaerobic conditions were conducted in 1D and 2D.

The overall biodegradation rate depends strongly on Henry's law constant. Highly volatile compounds are slowly degraded (see iso-octane, Fig. 3) compared to low H compounds such as BTEX (see toluene in Fig. 3), because biodegradation takes place only in the aqueous phase.

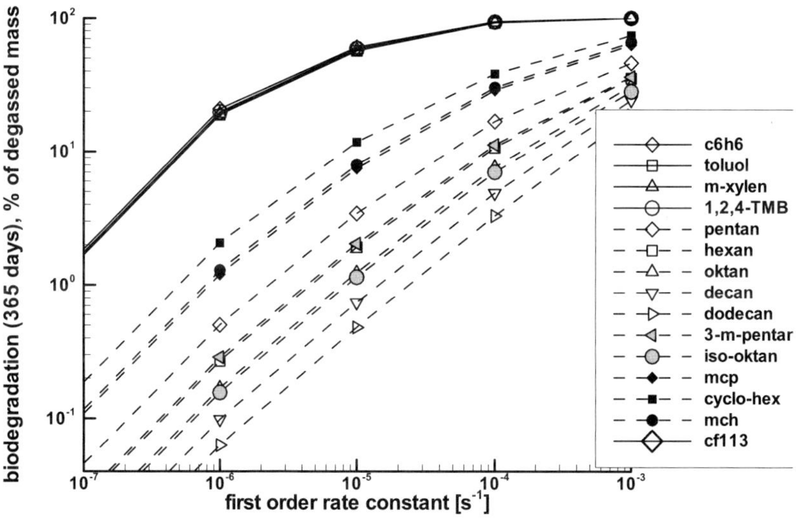


**Fig. 3.** Sensitivity of total biogegradation after one year on biodegration rate constant in terms of cumulated biodegradation in % of mass volatilized from the NAPL source. Compounds differ in Henry's law constant H: BTEX (upper four) about 0.2, alkanes (pentane to iso-octane) between 70 and 300 and lower four (cyclo-alkanes and tracer) between 5 and 20.

The ratio of the atmospheric escape to biodegraded mass versus H at a constant degradation rate constant (pseudo first order) is shown in Fig. 4. It demonstrates that low H compounds degrade more efficiently. This can be explained by the large fraction of such compounds in the aqueous phase.

Depletion of the NAPL phase is accelerated by rapid biodegradation as shown in Fig. 5. The mass still present after 1 year decreases with increasing degradation rate. This is because the evaporation of the NAPL constituents is enhanced due to steep concentration gradients which are maintained by biodegradation.

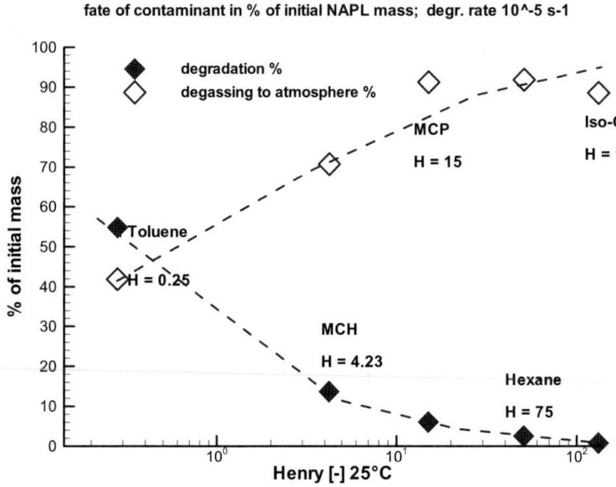


**Fig. 4.** Fate of organic contaminants in relation to Henry's Law constant. A uniform pseudo-first order biodegradation rate constant =10-5 s-1 was applied for all the contaminants. GRACOS- Værløse 2D scenario (14 compound mixture), depth of contaminant source 1 m. Henry's law constants from Pasteris et al. (2001).

Increasing soil water contents reduce the degassing to the atmosphere and thus slightly increase the fraction biodegraded as long as no oxygen limitation occurs. Higher temperatures enhance the volatilisation of the NAPL constituents in general and therefore lead to both, more rapid degassing to the atmosphere and faster biodegradation. For all the volatile compounds involved, groundwater recharge was found to have only minor influence on the amount of mass which was transported into the saturated zone. An increase in transverse vertical dispersivity in the capillary fringe region, on the other hand, significantly enhances transport to groundwater.

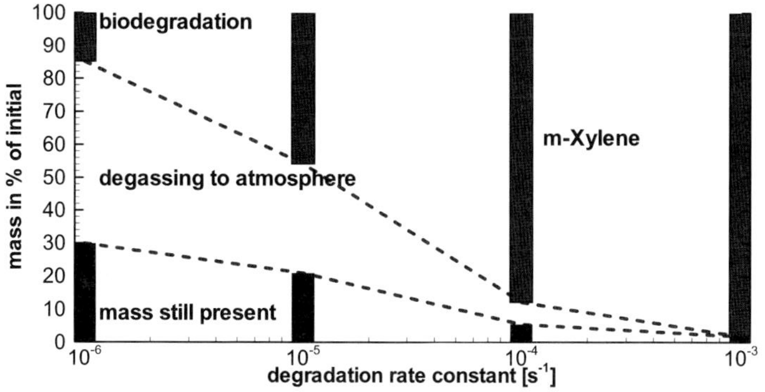


**Fig. 5.** Fate of m-xylene in relation to the pseudo first order biodegradation rate constant. GRACOS- Værløse 2D scenario (14 compound mixture), depth of contaminant source 1 m, elapsed time: 1 a.

The significance of oxygen depletion at the NAPL source and possible anaerobic degradation scenarios were also tested using the model. Three different anaerobic scenarios can be distinguished: (1) degradation by electron acceptors such as nitrate and sulphate, (2) reduction of iron oxide or manganese oxide containing soil minerals and (3) methanogenesis. Nitrate and sulphate reduction (1) can be neglected due to the low concentrations in rain (Scheffer and Schachtschnabel 1992). Reduction of oxides (2) can occur e.g. in iron oxide rich soils, but is kinetically limited. Methanogenesis (3) is often observed in heavily contaminated soils if the oxygen supply is restricted. The scenarios required to establish anaerobic conditions were evaluated further as described below.

In a 1D scenario there is no lateral possibility for oxygen supply and anaerobic conditions can develop quickly, but this scenario is not very realistic for a field case. In a 2D model, the values required for methanogenesis at the field site are constant volumetric water content larger than 14 % (with a porosity of 35 %, a benzene fermentation rate constant of 7.8 x 10-6 mg $l^{-1}$ $s^{-1}$ (zeroth order), a methane oxidation rate constant of 10-5 $s^{-1}$ (pseudo first order), and an inhibition threshold of 0.03 mg $l^{-1}$ oxygen). The NAPL source in this scenario contained only benzene, which is highly soluble and fast degassing, i.e. rapidly consuming oxygen (Fig. 6). Under this condition, small concentrations of methane (< 1 % Vol.) occur at the capillary fringe below the NAPL source. If soil water contents approach saturation, then the zone of methane formation moves closer to the NAPL source due to faster oxygen depletion and restricted diffusive supply of oxygen from the atmosphere.

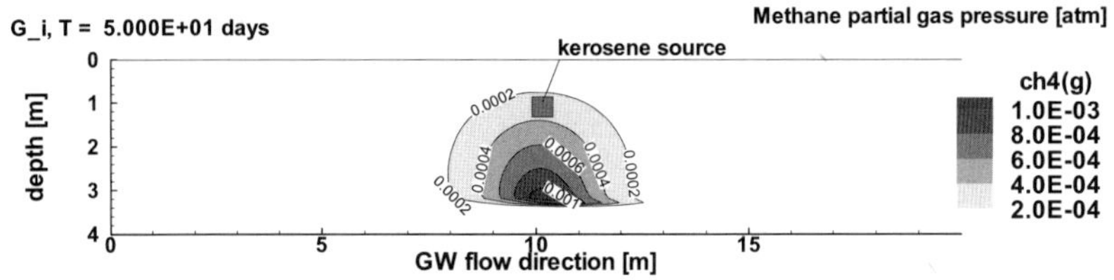


**Fig. 6.** Methane concentration at maximum, 20 days after source installation in a 2D simulation.

For the kerosene composition used in the field experiment, however, oxygen depletion was neither predicted by the model nor observed at the field site. Thus, anaerobic conditions can only be expected for fine grained material with high water content or very large, extended NAPL sources, especially if they are deep or close to the groundwater table.

## Model validation

The numerical simulations demonstrated that in the scenario investigated most of the contaminant mass in the unsaturated zone will either be transported to the atmosphere and be subject to biodegradation. Only a very small fraction will diffuse across the capillary fringe and end up in the groundwater. Parameters, which are not measured independently have to be fitted in the model. Those are most likely e.g. the biodegradation rate constants in the unsaturated zone. After the contaminant spreading in the vadose zone was quantified, the mass transfer to groundwater was evaluated by realistic estimates of transverse vertical dispersion and diffusion in the capillary fringe (see also Klenk and Grathwohl 2002, Eberhardt and Grathwohl 2002, Gaganis et al. 2002).

Potential parameters that allow for model validation by comparing them to measured field data are the mole fractions of the compounds present in the NAPL phase over time, as well as the vapour phase concentrations observed in the unsaturated zone. Simulated NAPL phase contents show very good agreement to the field data (Fig. 7).

The graph depicts that high volatile compounds like methyl-cyclo-pentane (MCP) are depleted early after about 100 days, in the meanwhile lower volatile components such as octane show a relative increase in their mole fraction until they eventually become depleted. Fig. 8 shows the good agreement of field observations with the numerical simulations accounting for the processes and parameters discussed above.

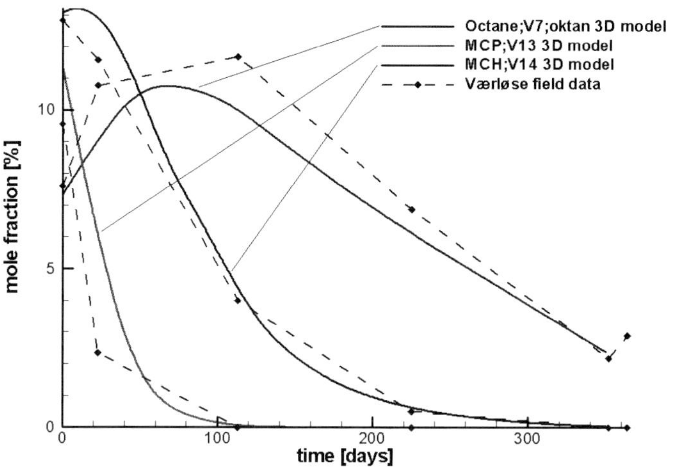


**Fig. 7.** Comparison of measured and simulated mole fractions within the NAPL source for the three selected kerosene compounds n-octane, methyl-cyclo-pentane (MCP) and methyl-cyclo-hexane (MCH).

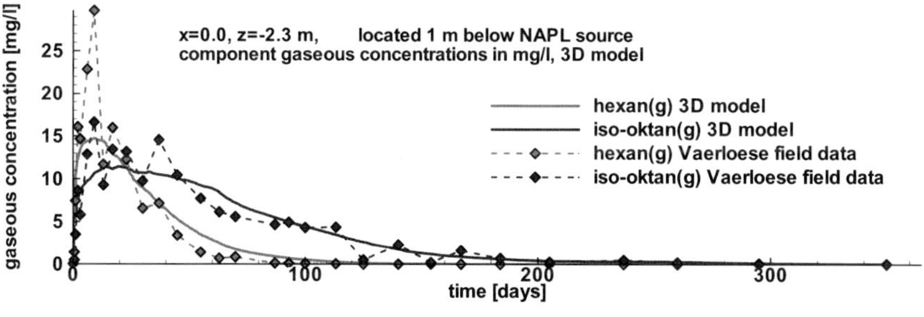


**Fig. 8.** Comparison of measured and simulated concentrations of two selected kerosene compounds (n-hexane and iso-octane) at a sampling port 1 m below the centre of the kerosene source.

If the model accounts for a depth and time-dependent temperature field (lower vapour pressures in the winter months), then the agreement with measured data improves significantly compared to constant temperature simulations. A temporally variable unsaturated flow field due to infiltration events results in elevated concentrations of CO2 during longer precipitation periods as shown in Fig. 9. This can be explained by higher pore water saturation in the top soil which diminishes the diffusive transport of the biodegradation product to the atmosphere.

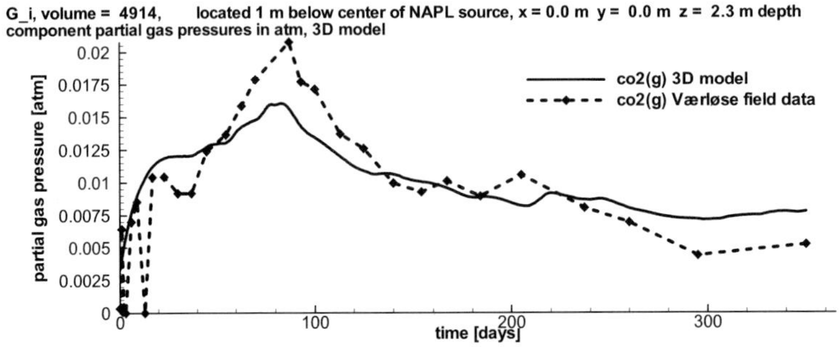


**Fig. 9.** Comparison of measured and simulated concentrations of the reaction product carbon dioxide at a sampling port 1 m below the centre of the kerosene source.

## Results of the field site model

As shown above, the contaminant spreading at the field site and the concentrations of 13 kerosene compounds and the tracer could be reproduced very well by the calibrated model. A view of a vertical cross-section through a 2D model domain is shown in Fig. 10 for the concentrations of iso-octane, and a time series of the vapour phase plume for the compound cyclo-pentane in Fig. 11.

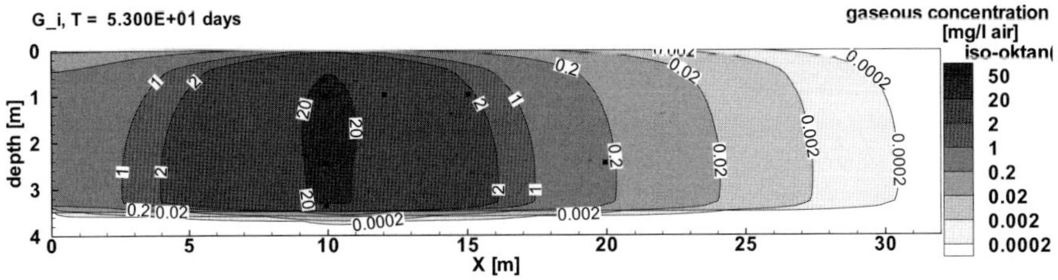


**Fig. 10.** Vertical cross section through the model domain for a 2D scenario and the vapour phase concentrations of iso-octane 53 days after source installation (time of maximum spreading of the soil gas plume).

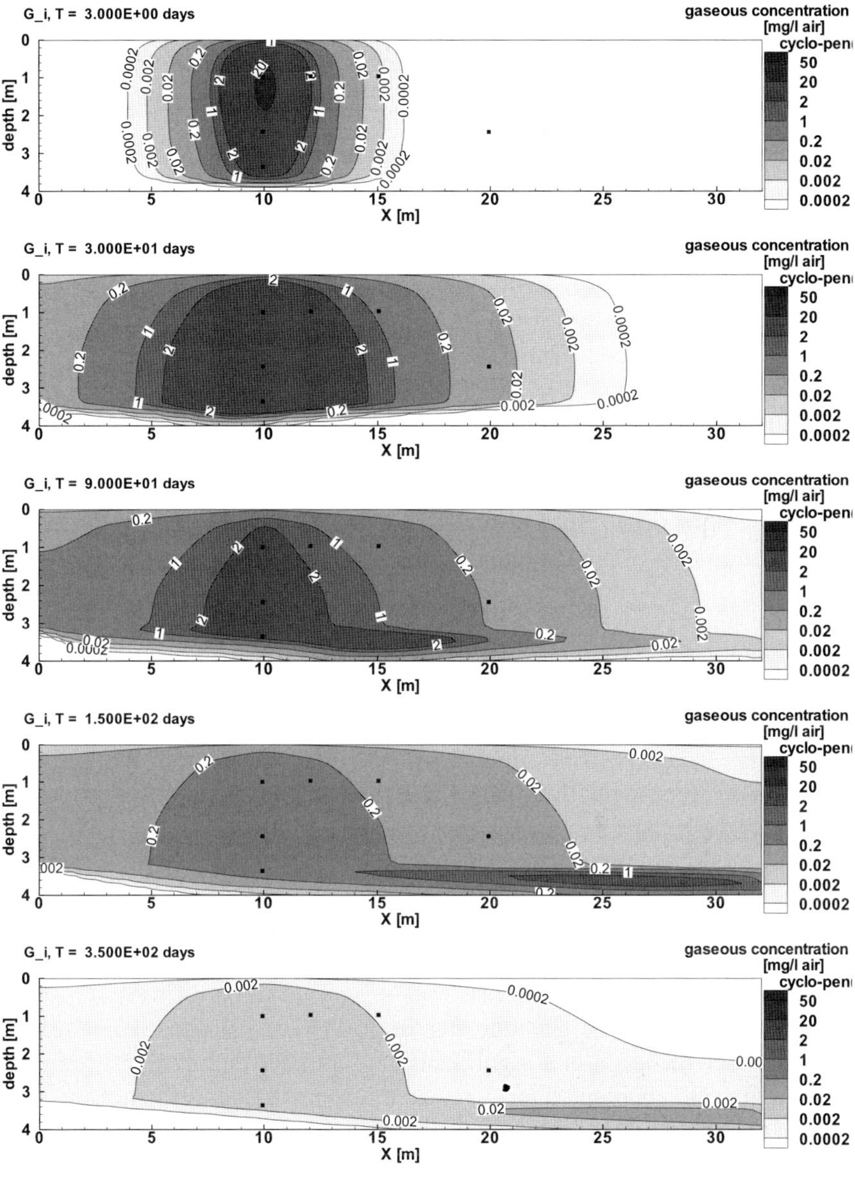


**Fig. 11.** Time series of Cyclo-Pentane concentration contours in the vapour phase, vertical cross section through the 2D model domain. Time: 3, 30, 90, 150 and 350 days after source installation.

Iso-octane is the most concentrated compound in the original NAPL. It can be seen that the plume is restricted to the unsaturated zone, only very low concentrations in groundwater and the capillary fringe region were observed. The concentration decreases rapidly from the source zone.

Cyclo-pentane is one of the highest volatile compounds in the artificial kerosene mixture with a Henry-constant of 7.6, significantly lower than that of iso-octane (123). Therefore it shows considerable biodegradation (~ 8 % of mass) and is transported into groundwater (0.4 % of mass). Significant contamination extends only in an area of 5 m in diameter due to biodegradation and volatilisation to the atmosphere. After an initial phase of fast spreading, the compound is depleted from the source after about 60 days but remains present in the subsurface for much longer. Transport to groundwater is delayed by slow diffusive and dispersive transport, until a significant shallow plume can be observed after 90 days which is dragged downgradient by groundwater flow. Due to volatilisation to the atmosphere (90 % of mass), the soil air plume has almost disappeared after the one year duration of the experiment.

Ageing of the NAPL source results in the enrichment of the lower volatile compounds on the expense of the components with higher vapour pressure, which escape to the atmosphere. The transient evolution of the composition of the NAPL source is shown in Fig. 12, which can be regarded as a typical example of kerosene ageing during one year after spill. After 350 days only 5 compounds remain in the residual phase in significant amounts: xylene, tri-methyl-benzene, n-octane, n-decane, n-dodecane and methyl-cyclo-hexane.

The contaminant flux into groundwater is very small compared to volatilisation of source constituents to the atmosphere. As already observed in previous studies (Klenk 2000, Pasteris et al. 2001) the results show that for VOCs most of contaminant mass will be degassing to the atmosphere or is subject to biodegradation, whereas only a small fraction diffuses across the capillary fringe into the groundwater, as long as the surface is not sealed. Groundwater recharge is a significant contaminant pathway only in soils with high water content due to reduced vapour phase diffusion, and for compounds with low Henry's Law constants, e.g. MTBE in silt or clay. Maximum transport rates into groundwater occurred for the 2 most volatile and soluble compounds toluene and cyclo-pentane. Even in this case only about 3% and 0.4%, respectively, of the total mass of each compound was transported into the groundwater. Concentrations measured in shallow groundwater could be reproduced best applying a vertical dispersivity $\alpha_t$ of 2 cm in the model.

The flux to the atmosphere was most pronounced for the tracer (freon 113), the light alkanes (n-, iso- and cyclo-alkanes) due to their high vapour pressure and low biodegradation rate constants. Biodegradation was most important for the aromatic compounds (low Henry's law constant) and the heavier n-alkanes, which showed high biodegradation rate constants. According to the model a total mass of 5 kg $O_2$ was consumed during one year of the experiment leading to production of $CO_2$ and water (oxidation of hydrogen contained in the hydrocarbons). No significant depletion of $O_2$, however, was observed even at the source location, which coincides well with the field measurements.

The contaminant spreading at the field site and the concentrations of 13 kerosene compounds and the tracer could be reproduced very well by the model calibrated to the biodegradation rate constants. It should be noted, however, that for a number of compounds which are not biodegraded (tracer CF113) or have only weak sensitivity on biodegradation (hexane, iso-octane, e.g.), the results come close to pure forward prediction solely based on field and compound parameters.

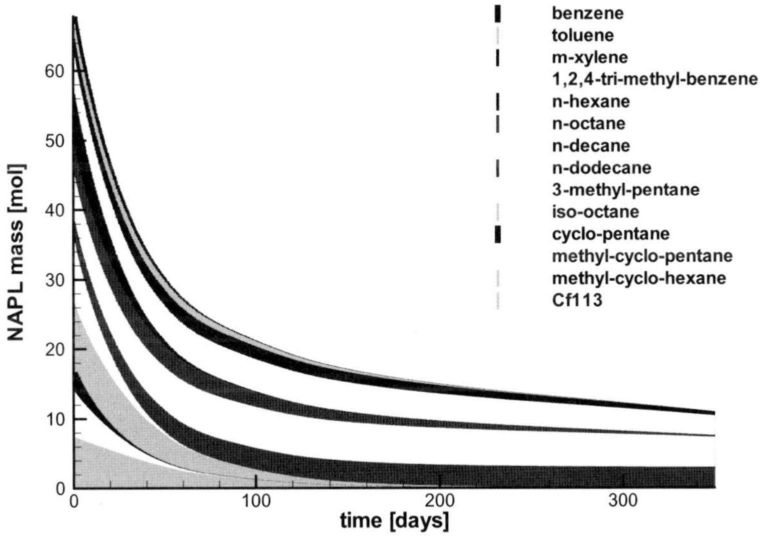


**Fig. 12.** Change of the kerosene composition during the first year as simulated in a 3D model. Sorted in compound classes, from top to bottom, initially: Benzene, Toluene, 1,2,4-Tri-Methyl-Benzene, Hexane, Octane, Decane, 3-Methyl-Pentane, iso-Octane, Cyclo-Pentane, Methyl-Cyclo-Pentane, Methyl-Cyclo-Hexane, and the tracer freon CF113 (only in very small fraction).

## Summary and conclusion

Measured data at the well controlled field experiment of a kerosene spill at Værløse field test site can be well predicted with the numerical model MIN3P. Volatilisation to the atmosphere proved as the most important Natural Attenuation process in the unsaturated zone for volatile organic pollutants from shallow spills, followed by biodegradation. Modelling results illustrate that the overall biodegradation rates depend mainly on distribution parameters such as Henry's Law constant of the fuel constituents, on the biological degradation rate constant, on the soil water content and temperature. The contaminant transfer into groundwater depends on all of these properties and can be determined after quantification of the processes in the unsaturated zone. For volatile compounds, transport to groundwater was found to be controlled by dispersion-diffusion processes. Results of the field investigations allow the validation of the model, which then can be used to quantify the risk of groundwater contamination for a variety of different scenarios.

## Acknowledgement

This work was supported by the EU 5$^{th}$ framework programm project GRACOS (Groundwater Risk Assessment at Contaminanted Sites, EVK1-CT1999-00029).

## References

Broholm MM, Christophersen M, Maier U, Stenby EH, Kjeldsen P (2005) Compositional evolution of the emplaced fuel source in the vadose zone field experiment at Airbase Værløse, Denmark. Environmental Science and Technology submitted.

Christophersen M, Broholm M, Kjeldsen K (2002) Migration and degradation of fuel vapours in the vadose zone. In Halm & Grathwohl (eds.): Groundwater Risk Assessment at Contaminated Sites. Proceedings of the 1$^{st}$ GRACOS workshop, Tübingen 22./23.02.02, 83-87

Christophersen M, Broholm M, Mosbæk H, Karapanagioti HK, Burganos VN, Kjeldsen P (2005) Transport of hydrocarbons from an emplaced fuel source experiment in the vadose zone at Airbase Værløse, Denmark. Journal of Contaminant Hydrology submitted

Eberhardt C, Grathwohl P (2002) Time scales of pollutants dissolution from complex organic mixtures: blobs and pools. J Cont Hydrol 59, 1-2, 45-66 (Invited paper for Special Issue of J Cont Hydrol on Site Remediation)

Gaganis P, Skouras ED, Theodoropoulou M, Burganos V (2002) Modelling contaminant emission across the capillary fringe. In Halm & Grathwohl (eds.): Groundwater Risk Assessment at Contaminated Sites. Proceedings of the 1st GRACOS workshop, Tübingen 22./23.02.02, 119-124

Grathwohl P, Klenk I (2000) Groundwater risk assessment at contaminated sites (GRACOS). In W. Harder et al., (eds). Contaminated Soil 2000 Vol.2:831-834, Leipzig: Thomas Telford

Höhener P, Dakhel N, Werner D, Pasteris G (2002) Biodegradation of volatile organic compounds in the unsaturated zone. In Halm & Grathwohl (eds.): Groundwater Risk Assessment at Contaminated Sites. Proceedings of the 1st GRACOS workshop, Tübingen 22./23.02.02, 88-93

Hölting B (1992). Hydrogeologie. Stuttgart, Enke, 415 p

Hughes-Conant B, Gillham RW, Mendoza CA (1996) Vapor transport of trichlorethylene in the unsaturated zone: Field and numerical modeling investigations. Water Resources Research vol.32 no. 1:9-22

Klenk ID (2000) Transport of volatile organic compounds (VOC's) from soil-gas to groundwater. PhD Dissertation, TGA C55, Center for Applied Geosciences, Tübingen

Klenk ID, Grathwohl P (2002) Transverse vertical dispersion in groundwater and the capillary fringe. J Cont Hydrology 58, 1-2, 111-128

Mayer KU (1999) A multicomponent reactive transport model for variably saturated media. PhD-Thesis at the University of Waterloo, Department of Earth Sciences, Waterloo, Ontario, Canada

Mayer KU, Frind EO, Blowes DW (2002) Multicomponent reactive transport modeling in variably saturated porous media using a generalized formulation for kinetically controlled reactions. Water Resources Research 38(9):1174-1195

Pasteris G, Werner D, Kaufmann K, Höhener P (2001) Vapor phase transport and biodegradation of volatile fuel compounds in the unsaturated zone: a large scale lysimeter study. Environ Sci Technol 36:30-39

Scheffer F, Schachtschabel P (1992) Lehrbuch der Bodenkunde. 13. Auflage, Stuttgart, 491 p

Wang G, Reckhorn SBF, Grathwohl P (2003) Volatilization of VOC from multicomponent mixtures in unsaturated porous media. Vadose Zone Journal 692-701

Wiedemeier TH, Rifai HS, Newell CJ, Wilson JT (1999) Natural Attenuation of Fuels and Chlorinated Solvents in the Subsurface. John Wiley and Sons, New York

# Enhancement of solute spreading in soils due to particle-facilitated transport and preferential flow

S. Bold*, R. Liedl, P. Grathwohl

Center for Applied Geoscience, University of Tübingen
* now at Emschergenossenschaft / Lippeverband

## Abstract

A process-based modelling approach of organic contaminant spreading is presented, which allows to quantify the increase of groundwater pollution risk due to preferential flow and particle-facilitated transport in the vadose zone. Both effects are illustrated by means of a parameter study based on realistic input data. When coupled with biodegradation, preferential flow and particle-facilitated transport lead to an increase in the long-term concentration to be expected at the groundwater table as both processes reduce the bioavailability of the contaminant due to either shorter residence times (preferential flow) or lower accessibility (particle-facilitated transport). Model predictions account for the amount and the velocity of preferential flow as well as the potentially kinetic interaction between organic contaminant and mobile particles.

## Keywords

Unsaturated zone; organic contaminants; groundwater pollution risk; process-based modelling

## Introduction

Point sources of organic contaminants in the unsaturated zone are a potential threat to groundwater quality. Groundwater risk assessment therefore strongly requires valid information about contaminant concentrations to be expected in the vadose zone water and at the unsaturated zone-groundwater interface, the latter representing an important point of compliance. The prediction of contaminant concentrations may be supported by contaminant transport modelling, which has to account for the relevant processes like sorption, biodegradation, preferential flow and particle-facilitated transport. In particular, it is of major importance to predict the interactions of processes which are responsible for an increase (preferential flow and particle-facilitated transport) or a decrease of groundwater pollution risk (sorption and biodegradation).

This paper is aimed to investigate the influence of preferential flow paths and kinetically controlled particle-facilitated transport on the transport behaviour of contaminants by means of scenario calculations. To this end, a process-based streamtube modelling approach is employed, which takes into account diffusion-limited sorption / desorption kinetics, biodegradation, particle-facilitated transport and preferential flow.

## Modelling Approach

The transport model used in this study is taken from Bold et al. (2003) who apply the Lagrangian framework presented by Dagan and Cvetkovic (1996) to describe reactive solute transport. This approach allows for an efficient separate treatment of conservative transport and reactive processes. A generic expression for the total mobile contaminant concentration at the point of interest (column outlet) is given by

$$C(t) = \int_0^\infty g(\tau)\Gamma(\tau,t)d\tau \tag{1}$$

where $t$ is time [T] and $\tau$ represents the travel time of a conservative tracer through the model domain [T]. To account for variations in arrival times, e.g. due to hydraulic heterogeneities, a probability density function (PDF) of travel times $g(\tau)$ is used. Several theoretical expressions for $g(\tau)$ are given in Jury and Roth (1990). Within this concept, preferential flow can be easily incorporated by using a bimodal PDF (Utermann et al. 1990). The function $\Gamma(\tau,t)$, which has been termed reaction function by Dagan and Cvetkovic (1996), quantifies reactive processes for a continuous injection with unit input concentration.

In order to overcome the restriction of analytical expressions for $g(\tau)$ and $\Gamma(\tau,t)$, the integral in eq. (1) is replaced by a finite sum yielding $C(t) = \sum_{l=1}^{n} g(\tau_l)\Gamma(\tau_l,t)\Delta\tau$ where $\Delta\tau$ is the discretisation of the one-dimensional streamtube model with respect to travel time $\tau$ and $l$ is the time step index. Based on this relationship, the model SMART (Streamtube Model for Advective and Reactive Transport) developed by Finkel et al. (1999) is applied in this study. The underlying process model is described in the following sections.

## Particle and Contaminant Transport

Neglecting the influence of immobilised particles on the effective porosity, particle transport along a streamtube, which is parametrised with respect to travel time, is quantified by

$$\frac{\partial}{\partial t}C_c + \frac{\partial}{\partial \tau}C_c = -KC_c \tag{2}$$

where $C_c$ is the mobile particle concentration [$ML^{-3}$] and $K$ is the particle deposition rate coefficient [$T^{-1}$]. The right-hand side of eq. (2) indicates that particle deposition is modelled by a first-order rate law.

Contaminants can sorb in the matrix and on mobile and immobile particles. Taking this into account, contaminant transport is described by

$$\frac{\partial}{\partial t}\left[C_D + \frac{\rho_b S_{matrix}(C_D)}{n_e} + S_{cm}(C_D)C_c + \frac{S_{cim}(C_D)\rho_c\sigma_c}{n_e}\right] + \\ + \frac{\partial}{\partial \tau}\left[C_D + S_{cm}(C_D)C_c\right] = -\lambda C_D \tag{3}$$

where $C_D$ is the dissolved contaminant concentration [$ML^{-3}$], $\rho_b$ is the bulk density of the porous medium [$ML^{-3}$], $\rho_c$ is the particle dry solid density, $\sigma_c$ is the volume of immobile particles per unit volume [-] and $\lambda$ is the first-order degradation rate constant [$T^{-1}$]. $S_{matrix}(C_D)$, $S_{cm}(C_D)$, $S_{cim}(C_D)$ are the ratios of sorbed solute mass per unit mass of solid matrix, mobile and immobile particles, respectively. The relationships $S(C_D)$ depend on the mechanisms being responsible for the contaminant-solid interactions.

If only the total mobile contaminant concentration $C_{INP}$ entering the modelling domain is known instead of dissolved contaminant concentration $C_{D,INP}$, the corresponding boundary condition for eq. (3) can be calculated by $C_{D,INP} = \beta C_{INP}$ where $\beta$ is the contaminant mass ratio which is not subject to particle-facilitated transport [-]. Assuming that contaminant and mobile particles entering the modelling domain are pre-equilibrated and neglecting the volume of the intra-particle pores of the particles, $C_{D,INP}$ can be quantified by solving $C_{D,INP} + S_{cm}(C_{D,INP})\, C_{c,INP} = C_{INP}$ where $C_{c,INP}$ is the concentration of mobile particles entering the model domain [$ML^{-3}$].

## The Intra-particle Diffusion Model

The intra-particle diffusion model is used to simulate both, the interactions of contaminant with the matrix and with mobile and immobile particles (Figure 1). Within this approach, the sorption / desorption behaviour of organic chemicals is regarded as a kinetic process, limited by the diffusive transport of the solute through the tortuous pores of the matrix or particles to the sorption sites, whereas the actual process of adsorption or desorption of a molecule at the sorption site is comparatively fast and can be described as equilibrium reaction.

As the matrix grains / particles are assumed to be spherically symmetric, the underlying intra-particle diffusion equation – based on Fick's 2nd law – can be written for a specific matrix grain / particle as

$$\frac{\partial}{\partial t}[\varepsilon c + (1-\varepsilon)\rho\ s(c)] = D_{eff}\frac{1}{r^2}\frac{\partial}{\partial r}(r^2\frac{\partial c}{\partial r}) \tag{4}$$

with the initial and boundary conditions $\partial c/\partial r(\tau, r=0, t) = 0$ and $c(\tau, r=R, t) = C_D(\tau, t)$ where $r$ is the radial coordinate (distance from the centre of the sphere) [L], $D_{eff}$ is the effective diffusion coefficient [$L^2T^{-1}$], $\varepsilon$ is the intra-particle porosity [-], $\rho$ is the dry solid density of matrix grains / particles, $c$ is the dissolved contaminant concentration in the intra-particle pores, $R$ is the grain / particle radius and $s$ is the mass of chemical sorbed onto surfaces of intra-particle pores per unit mass of the grains / particles. $s = s(c)$ may represent any type of sorption isotherm, e.g. linear, Freundlich, Langmuir or others.

According to Grathwohl (1998), the effective diffusion coefficient $D_{eff}$, accounting for the reduction of diffusion due to the tortuous pores and reduced diffusion-effective area, can be empirically estimated by $D_{eff} = D_{aq}\varepsilon^m$ where $D_{aq}$ is the diffusion coefficient in aqueous solution [$L^2T^{-1}$] and $m$ is an empirical exponent [-] found to be close to 2 for many porous rocks (Rügner et al. 1999, Boving and Grathwohl 2001).

From eq. (4) the mass of the contaminant within all grains of the matrix $S_{matrix}$ and within the particles $S_{cm}$ and $S_{cim}$ per unit volume can be obtained, which is required for solving the reactive transport equation (3). According to Liedl and Ptak (2003), the contaminant mass is given by

$$\begin{aligned} S_{matrix}(C_D) &= S_{cm}(C_D) = S_{cim}(C_D) = \\ &= \frac{1}{\frac{4\pi}{3}R^3(1-\varepsilon)\rho}\int_0^R 4\pi r^2\left[\varepsilon c + (1-\varepsilon)\rho s(c)\right]dr \end{aligned} \tag{5}$$

where the right-hand side is to be evaluated for the matrix material or the particles, respectively.

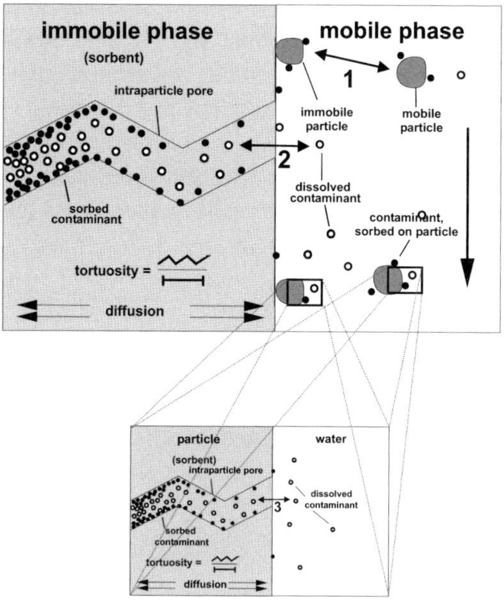


**Fig. 1.** SMART concept of particle-facilitated transport using the intra-particle diffusion model (Bold et. al 2003)

## Equilibrium vs. Decoupled Transport

If the sorption rate of the solute is significantly larger than its mass flow rate, sorption / desorption of a contaminant onto / from mobile particles can be regarded as an equilibrium reaction. This behaviour may be hypothesised for small particles and will be investigated in Section 5.

If the equilibrium approximation is valid, $S_{cm}(C_D)$ and $S_{cim}(C_D)$ in eq. (4) can be obtained by $S_{cm}(C_D) = S_{cim}(C_D) = s(C_D)$ where $s(C_D)$ represents any type of sorption isotherm between dissolved contaminant and particles.

In contrast, if solute mass transfer between mobile particles and the dissolved phase is much slower than advective mass transport, particle-facilitated transport of contaminants can be regarded as fully decoupled from the dissolved phase and the contaminant transport equation reads

$$\frac{\partial}{\partial t}\left[C_D + \frac{\rho_b S_{matrix}(C_D)}{n_e}\right] + \frac{\partial C_D}{\partial \tau} = -\lambda C_D \qquad (6)$$

The Damköhler number (e.g., Jennings and Kirkner 1984) can be used to determine whether a reversible chemical reaction is fast enough to assume equilibrium or slow enough to neglect any sorption / desorption of the particle-associated contaminants. The Damköhler number is defined as the ratio between the transport and the reaction time scales. Within the Lagrangian framework it is given by $D_a = \lambda_{diff}\bar{\tau}$ where $\lambda_{diff}$ is the rate constant for the diffusion-limited mass transfer [$T^{-1}$] and $\bar{\tau}$ is the mean travel time of the particles [T], i.e. $\bar{\tau} = \int_0^\infty \tau g(\tau) d\tau$. The rate constant $\lambda_{diff}$ is proportional to the corresponding diffusion rate constant. To allow for an analytical definition of this quantity, it is assumed that equilibrium sorption of contaminants in the intra-particle pores of mobile particles can be described by a linear isotherm. For nonlinear isotherms, a representative linear isotherm may be used as an approximation. Using this assumption and employing an analytical solution of eq. (4) (Crank 1975), the reaction rate constant can be quantified as a function of $\bar{\tau}$ by (Grathwohl 1998)

$$\lambda_{diff} = -\ln\left[\frac{6}{\pi^2}\sum_{n=1}^{\infty}\frac{1}{n^2}\exp\left(-n^2\pi^2\bar{\tau}\frac{D_{app}}{R^2}\right)\right]/\bar{\tau} \qquad (7)$$

where $D_{app}$ is the corresponding apparent diffusion coefficient [$L^2T^{-1}$], which in case of porous mobile particles can be calculated by $D_{app} = D_{eff}/(\varepsilon + K_{d,c}(1-\varepsilon)\rho)$ with $K_{d,c}$ denoting the partitioning coefficient between contaminant and mobile particles [$L^3M^{-1}$].

## Basic Scenario

In order to investigate the effect of preferential flow and particle-facilitated transport on the transport behaviour of organic contaminants, contaminant breakthrough curves (BTCs) at the point of compliance are calculated for several realistic scenarios. The basic scenario is outlined in Figure 2 with model parameters given in Table 1.

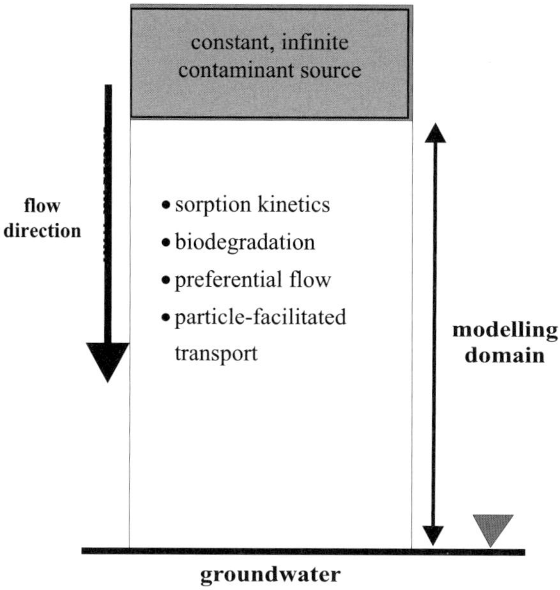


**Fig. 2.** Basic scenario

**Table 1.** Model parameters for basic scenario (Figure 2)

| parameter | unit | value or range |
|---|---|---|
| inflow concentration | mg/L | 10.0 |
| mean travel time (matrix) | a | 1.0 |
| mean travel time (pref. flow) | a | 0.1 – 1.0 |
| amount of pref. flow | % | 0 – 50 |
| water content | - | 0.15 |
| bulk density | $kg/m^3$ | $1.90 \cdot 10^{-3}$ |
| solid density | $kg/m^3$ | $2.73 \cdot 10^{-3}$ |
| grain diameter | mm | 2.0 |
| intra-particle porosity | - | 0.01 |
| distribution coefficient (immobile solids) | L/kg | 1.0 |
| aqueous diffusion coefficient | $cm^2/s$ | $7.68 \cdot 10^{-6}$ |
| half-life | a | 1.0 |
| particle concentration | mg/L | 0.0 – 10.0 |
| partitioning coefficient (mobile solids) | L/kg | 10000 |

## The Effect of Preferential Flow

For the general scenario described above, contaminant BTCs are calculated with SMART for several bimodal PDFs (Figure 3). For the scenario calculations both the amount and the mean travel time of water entering the preferential flow region is varied. Thereby, particle-facilitated transport is neglected.

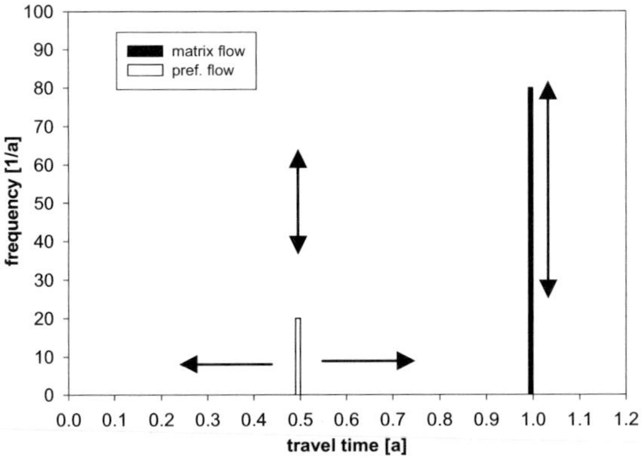


**Fig. 3.** Bimodal PDF influenced by preferential flow

In Figure 4, contaminant BTCs influenced by preferential flow are shown. Thereby, in Figure 4a, the effect of the travel time of the preferential flow component is investigated assuming 10 % preferential flow. In order to observe an effect, the preferential flow velocity must be more than 10 % faster ($\tau_{pref} = 0.9$ a) than the matrix flow velocity. With decreasing travel time, the time of first arrival of contaminant at the control plane is decreasing. In addition, preferential flow results in an increasing long-term contaminant concentration. Similar results can be observed by varying the amount of preferential flow assuming that water entering the preferential flow region reaches the groundwater table after 0.5 a (Figure 4b).

The increasing long-term contaminant concentration due to preferential flow can be explained by a decreasing mean travel time of contaminant within the modelling domain, which results in reduced biodegradation. Therefore, by neglecting preferential flow, an underestimation of the long-term contaminant concentrations is introduced.

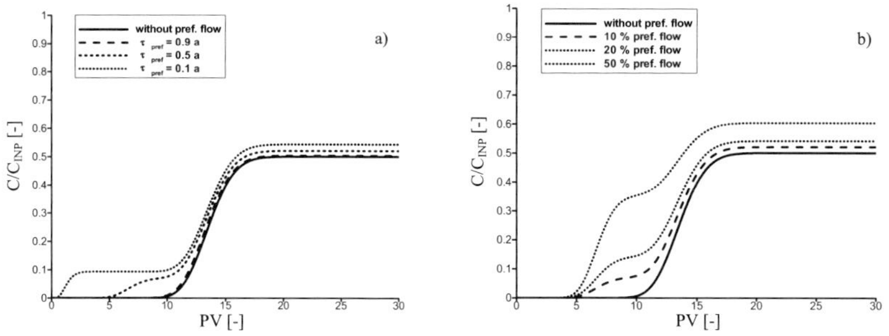


**Fig. 4.** Effect of preferential flow paths on contaminant BTC at the point of compliance assuming a constant contamination source: a) effect of travel time of preferential flow, b) effect of amount of preferential flow

## The Effect of Particle-facilitated Transport

In order to investigate the relative importance of sorption kinetics for modelling particle-facilitated transport, three approaches are used to describe contaminant-particle interaction: the intra-particle diffusion model taking into account kinetics and two simplified models, assuming equilibrium interaction and neglecting any interaction (decoupled model) between contaminant and particles. According to the time scale of sorption / desorption reactions, the equilibrium and the decoupled model can be viewed as end members for the model approach spectrum.

If not stated otherwise, simulations presented in this section rely on the following assumptions: (i) dispersion and preferential flow can be neglected, (ii) particles do not interact with the immobile soil matrix, and (iii) in the influent, the interaction of contaminant with particles is assumed to be in physico-chemical equilibrium described by a linear sorption isotherm with a distribution coefficient $K_{d,c}$ [$L^3M^{-1}$].

In Figure 5, the BTCs of a hydrophobic contaminant for the basic scenario are diagrammed for the three modelling approaches describing contaminant-particle interaction. The vertical axis represents the total concentration of mobile contaminants, $C = C_D + S_{cm}(C_D)C_c$, normalised by the total contaminant input concentration, $C_{INP}$. Without mobile particles, contaminant transport is only influenced by contaminant-matrix interaction and biodegradation resulting in a retarded solute spreading (crosses) which is affected by kinctically limited sorption / desorption of contaminant on the immobile solid phase. With mobile particles, quantifying contaminant-particle interaction as physico-chemical equilibrium process, this retardation is reduced (triangles). For a linear sorption isotherm the decrease of the retardation factor can be assessed by a factor of $1/(1+K_{d,c}C_c)$ if, in addition, physico-chemical equilibrium between contaminant and matrix is assumed (e.g., Magee et al. 1991). Neglecting contaminant-particle interaction (decoupled transport), the BTC can be seen as a superposition of retarded transport of dissolved contaminant and non-retarded transport of contaminant bound to particles (circles). Therefore, contaminants can be observed at the point of compliance after one pore volume, due to particle-facilitated transport (provided that the particles undergo conservative transport). Using the intra-particle diffusion model to take into account kinetic contaminant-particle interaction, BTCs highly depend on the Damköhler number, which provides a measure for the relative importance of reaction rates compared to the bulk flow rate (Jennings and Kirkner 1984). For a high Damköhler number ($D_a$ = 500, solid curve) representing "fast" kinetics the BTC is very similar to the BTC assuming equilibrium reaction, whereas for a small value ($D_a$ = 0.005, dotted curve), which is indicative for "slow" kinetics, little differences to the decoupled model can be observed. Only for $D_a$ = 1 (dashed curve) discrepancies to both, the equilibrium approach and the decoupled model are obvious.

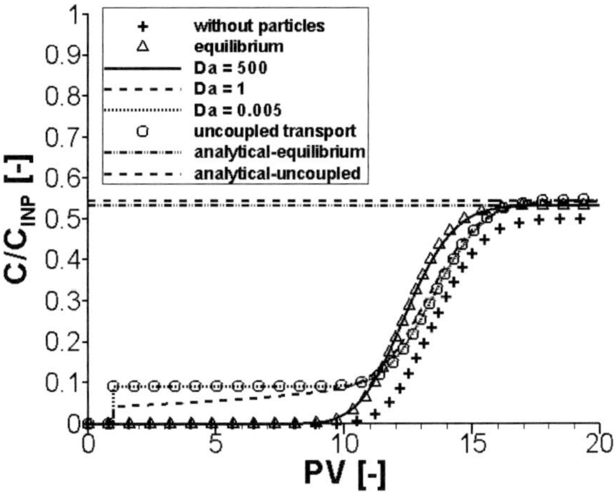


**Fig. 5.** Effect of particle-facilitated transport on contaminant BTC for different modelling approaches

In addition, particles cause a reduction in biodegradation indicated by higher long-term contaminant concentrations. Moreover, the long-term contaminant concentration is lower for "fast" sorption kinetics describing contaminant-particle interaction. This can be explained by noting that degradation of dissolved contaminants results in a desorption of contaminants from the particles. The desorbed contaminant is then again available for degradation by microorganisms. On the other hand, assuming no interaction between contaminants dissolved in water and contaminants sorbed onto the particles, a degradation of dissolved contaminants does not stimulate desorption in the uncoupled model.

The horizontal dash-dotted lines in Figure 5 represent the result of analytical expressions for the normalised long-term concentration of mobile contaminant, $C_\infty/C_{INP}$. Following Bold (2004), these relationships are given by

$$\frac{C_\infty}{C_{INP}} = e^{-\frac{\lambda \bar{\tau}}{1+K_{d,c}C_c}} \tag{8}$$

for the equilibrium scenario ($D_a = +\infty$) and by

$$\frac{C_\infty}{C_{INP}} = 1 - \beta + \beta e^{-\lambda \bar{\tau}} \tag{9}$$

for fully uncoupled transport ($D_a = 0$). In addition, Bold (2004) carried out a series of numerical simulations for Damköhler numbers between approximately $10^{-4}$ and $10^{6}$ in order to quantify the relative deviations of the long-term concentration from the equilibrium model (8) as a function of $D_a$ (symbols in Figure 6). Equivalently, ordinate values may be interpreted as the relative importance of uncoupled transport.

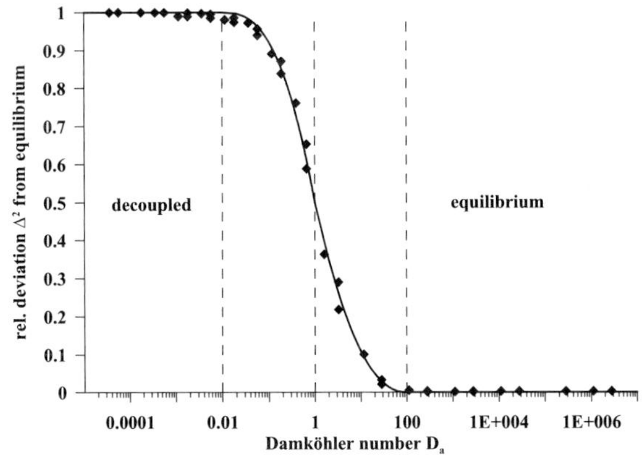


**Fig. 6.** Relative importance of uncoupled particle-facilitated transport as function of the Damköhler number. Symbols denote SMART model results obtained by Bold (2004)

Using these results, long-term concentrations of mobile contaminant can be determined for any Damköhler number by a simple arithmetic average of eqs. (8) and (9), i.e.

$$\frac{C_\infty}{C_{INP}} = \left[1-\Delta^2(D_a)\right]\cdot e^{-\frac{\lambda\bar{\tau}}{1+K_{d,c}C_c}} + \Delta^2(D_a)\cdot\left[1-\beta+\beta e^{-\lambda\bar{\tau}}\right] \tag{10}$$

where the weighting factor $\Delta^2(D_a)$ quantifies the relative importance of the uncoupled transport. By fitting power laws to the data (solid curve in Figure 6), $\Delta^2(D_a)$ can be fairly well approximated by

$$\Delta^2(D_a) = \begin{cases} 1-\frac{1}{2}\left(1+\frac{1}{2}\log D_a\right)^{2.69} & \text{if} \quad 0.01 < D_a \le 1 \\ \frac{1}{2}\left(1-\frac{1}{2}\log D_a\right)^{2.20} & \text{if} \quad 1 \le D_a < 100 \end{cases} \tag{11}$$

Furthermore, the weighting factor $\Delta^2(D_a)$ equals 1 for $D_a \le 0.01$ and 0 for $D_a \ge 100$, thus accounting for fully uncoupled and equilibrium transport, respectively, which constitute the end members of the spectrum of kinetic contaminant uptake / release by mobile particles.

As a result, eqs. (10) and (11) allow to straightforwardly use simple analytical expressions in order to assess long-term groundwater pollution including the effect of particle-facilitated transport with kinetic contaminant-particle interaction. Numerical simulations may therefore be avoided if the conditions (i) – (iii), which are mentioned at the beginning of this section, prevail.

## Conclusions

In order to predict the risk of groundwater pollution by organic contaminants, a numerical model was developed which takes into account sorption kinetics, biodegradation, particle-facilitated transport and preferential flow. This model is used to investigate the effect of preferential flow paths and kinetically controlled particle-facilitated transport on the transport behaviour of organic contaminants. Scenario calculations show that preferential flow paths and particle-facilitated transport not only accelerate solute spreading but also reduce biodegradation and therefore increase the long-term contaminant concentrations at the point of compliance. In many cases this increase may be minor, but if it becomes significant, then it is crucial to implement such processes in reactive transport models when assessing the groundwater risk due to potential pollution with organic contaminants.

## References

Bold S (2004) Process-based prediction of the long-term risk of groundwater pollution by organic non-volatile contaminants. Tübinger Geowissenschaftliche Arbeiten (TGA) C72

Bold S, Kraft S, Liedl R, Grathwohl P (2003) Sorption / desorption kinetics of contaminants on mobile particles: Modeling and experimental evidence. Water Resour Res 39(12):doi:10.1029/2002WR001798

Boving T, Grathwohl P (2001) Matrix diffusion coefficients in sandstones and limestones: Relationship to permeability and porosity. J Cont Hydrol 53(1-2):85-100

Crank J (1975) The mathematics of diffusion. Oxford University Press, Oxford

Dagan G, Cvetkovic V (1996) Reactive transport and immiscible flow in geological media. I General theory Proc R Soc London A 452:285-301

Finkel M, Liedl R, Teutsch G (1999) Modelling surfactant-enhanced remediation of polycyclic aromatic hydrocarbons. Environmental Modelling & Software 14:203-211

Grathwohl P (1998) Diffusion in natural porous media: Contaminant transport, sorption / desorption and dissolution kinetics. Kluwer Academic Publishers, Boston

Jennings AA, Kirkner DJ (1984) Instantaneous equilibrium approximation analysis J Hydraul Eng 110(12):1700-1717

Jury WA, Roth K (1990) Transfer functions and solute movement through soils. Birkhäuser, Basel

Liedl R, Ptak T (2003) Modelling of diffusion-limited retardation of contaminants in hydraulically and lithologically nonuniform media. J Cont Hydrol 66:239-259

Magee BR, Lion LW, Lemley AT (1991) Transport of dissolved organic macromolecules and their effect on the transport of phenanthrene in porous media. Environ Sci Technol 25(2):323-331

Rügner H, Kleineidam S, Grathwohl P (1999) Long-term sorption kinetics of phenanthrene in aquifer materials. Environ Sci Technol 33(10):1645-1651

Utermann J, Kladivko E, Jury WA (1990) Evaluating pesticide migration in tile-drained soils with a transfer function model. J Environ Qual 19:707-714

# Multiphase and Multi-component Interactions through the Unsaturated Saturated Zone Field and Model Study

S. Sorek[1, 2], M. Kuznetsov[1], A. Yakirevich[1], D. Ronen[1,3]

[1] Ben-Gurion University of the Negev, J. Blaustein Institutes for Desert Research, Zuckerberg Institute for Water Research, Environmental Hydrology & Microbiology, Sde Boker Campus, 84990, Israel
[2] Ben-Gurion University of the Negev, Mechanical Engineering, Pearlstone Center for Aeronautical Engineering Studies, Beer Sheva, 84105, Israel
[3] Water Quality Division, Israel Water Commission, POB 20365, Tel-Aviv, 61203, Israel

Corresponding author: Midreset Ben-Gurion, Sde Boker Campus 84990 Israel – Tel: +972 8 6596902 – Fax: +972 8 6596909 – e-mail: sorek@bgumail.bgu.ac.il

## Abstract

An integrated approach of field and model investigations was implemented to an aquifer underlying urban infrastructure. The study focuses on the transport of Trichloroethylene (TCE) through the unsaturated-saturated zone. Simulations, subject to isothermal conditions, addressed a three-dimensional continuum involving three interacting mobile phases: aqueous, NAPL (Non-Aqueous Phase Liquids) and gaseous. The mathematical model considers water, air and TCE as components in equilibrium partitioning between the three mobile fluids, while for the latter we account also for sorption on the solid matrix. Predictions of migration patterns were due to a continuous spill from a single NAPL source situated at the soil surface.

The specific flux of the gaseous phase proved to be directed to the surface at the vicinity of impervious segments, implying that TCE vapors may surround building foundations. We note a vertical, gravity driven, NAPL stem penetrating through the unsaturated-saturated zones. This causes a vertical displacement of the gaseous phase the velocity of which is also influenced by the gaseous pressure gradient. Fluids densities depend on TCE concentration, we thus note a density driven flow pattern. As TCE concentration decreases away from the NAPL stem, the gradient of the gas density yields its horizontal displacement, and vertical outward shift resulting from the vanishing capillary pressure. The gaseous velocity far away from the stem is associated only with the gas pressure gradient and is diminished at the unsaturated-saturated boundary, as the relative permeability of gaseous phase decreases near the capillary fringe.

## Keywords

Multiphase; multi-component; urban; modeling

## Introduction

The objective of the study was to examine by means of numerical simulations and available field observations, the temporal and spatial distribution of TCE resulting from leakage of NAPL at a prescribed location in the unsaturated and saturated zones of the Tel Aviv area. To achieve this goal, a three-dimensional (3-D) mathematical model of three $\ell$ phases and three $p$ components *l(≡ n - NAPL, q - aquenous; g - gaseous), p(≡ w water; a – air;* $c_m$ *– contaminants)* was used. The flow and transport of this multiphase and multi-component problem is a coupled mechanism, as explained in the following. The *linear momentum balance equation (Darcy's law)* for each $l$ phase reads

$$\mathbf{V}_l = -\mathbf{K}\frac{k_{rl}}{\mu_l}\left(\nabla P_l + \rho_l g \nabla Z\right), \quad \forall l \tag{1}$$

where $\mathbf{V}_l\,[m\,s^{-1}]$ denotes the $l$ phase velocity vector, $\mathbf{K}[m^2]$ denotes the absolute permeability tensor, $k_{rl}$ denotes the relative permeability of the $l$ phase, $\mu_l[Pa\,s]$ denotes the $l$ phase viscosity, $g[ms^{-2}]$ denotes the gravitational acceleration, $Z[m]$ denotes the altitude above some level and $\rho_l[kg\,m^{-3}]$ denotes the $l$ phase mass density. The latter suggests a density driven flow as its *state function* is defined by

$$\rho_l = M_l \sum_p W_p X_{pl}; \qquad \sum_p X_{pl} = 1, \ \forall l, p \tag{2}$$

where $W_p[kg\ mole^{-1}]$ denotes the molecular weight of the $p$ component, $M_l[mole\ m^{-3}]$ denotes the $l$ phase molar density and $X_{pl}$ denotes the mole fraction of the $p$ component in the $l$ phase. The *state functions* of $X_{pl}$ are associated with $P_l[Pa]$ the $l$ phase pressure, reading

$$X_{c_m g} = \frac{X^*_{c_m g} P^*_g}{P_g} X_{c_m n}; \quad X_{c_m g} = \frac{X^*_{c_m g} P^*_g}{X^*_{c_m q} P_g} X_{c_m q}; \quad X_{wg} = \frac{\beta_{wgq}}{P_g} X_{wq}$$
$$M_g = \frac{P_g}{RT^*}; \quad M_l / M_l^* = 1 + \chi_l \left(P_l - P_l^*\right), \quad l \equiv q,n \tag{3}$$

in which $P_l^*[Pa]$ denotes the $l$ phase reference pressure, $X^*_{c_m l}$ denotes the mole fraction of TCE in equilibrium with the $l \equiv g,q$ phases and $\beta_{wgq}[Pa]$ denotes the partial pressure of the water component in the gaseous phase which is in equilibrium with the aqueous phase, $M_l^*$ denotes the $l$ phase molar density at the $P_l^*$ pressure, $R[Pa\ m^3\ mole^{-1}\ {}^\circ K]$ denotes the ideal gas constant and $T^*[{}^\circ K]$ denotes a reference temperature and $\chi_l[Pa^{-1}]$ denotes the $l$ phase pressure compressibility. The phase pressures are related through the capillary pressures $(P_{c l_1 l_2})$ in the form

$$P_g = P_n + \alpha P_{cgn}(S_g) + \left(1-\alpha\right)\left[P_{cgq}(S_g) - P_{cnq}(S_q = 1)\right]; \qquad \sum_l S_l = 1,\ \forall l;$$
$$P_n = P_q + \alpha P_{cnq}(S_q) + \left(1-\alpha\right) P_{cnq}(S_q = 1) \tag{4}$$

where $S_l$ denotes the saturation of the $l$ phase, $\alpha = \min\left(1, S_n / S_n^*\right)$ and $S_n^*$ denotes a blending parameter associated with the NAPL phase.

Simulations were performed with the COMPFLOW code (Unger et al. 1996).

The actual contamination problem was solved over a total area of about 65.6 km2. The study area is part of the Coastal Plain aquifer of Israel which is mainly composed of layers of Pleistocene sand, sandstone and silt with intercalations of clay and loam. Some of the intercalations are discontinuous lenses; others near the coast are continuous and separate the aquifer into sub-aquifers, the upper is phreatic while the others are confined or partially confined towards the sea. The aquifer has 3-D lithological heterogeneity both at a macro and micro scale level (Ronen et al. 2000). To incorporate into the model the geological structure, 5 available hydrogeological cross-sections about 2000 m apart were digitized and schematized.

At the study area the depth of the domain is around 155 m and 75 m at the western and eastern borders, respectively, parallel to the sea shore. The groundwater level varies from 0 to 1 m above Mean Sea Level (MSL) at the western border and from 12 to 15 m at the eastern border. As a result of pumping, a groundwater depression was developed in a sub-area of the study domain, during the 1950s. The thickness of the unsaturated zone at the simulated NAPL spill location is about 20 m.

## Up-scaling

Multiphase simulators are effective tools for the investigation of complex processes operative in aquifers contaminated by volatile organic contaminants like TCE emerging from spills of NAPL. We followed the model of Forsyth et al. (1988, 1989), Slough et al. (1996, 1999) and Unger et al. (1996, 1998), and studied different operation scenarios. For the sake of comparison of observations at different spatial scales and to scale-up model predictions, we rewrite the mathematical model in a nondimensional form.

We will substitute each $\eta_{ij}$ scalar, vector or tensor quantity with $\tilde{\eta}\eta'_{ij}$, where $\tilde{\eta}$ and $\eta'_{ij}$ denote, respectively, the characteristic and dimensionless quantities of $\eta_{ij}$. Accounting for equilibrium partitioning of components between the phases, linear adsorption of TCE and isothermal conditions, we write (using Einstein's summation rule concerning repetitive indices) respectively, the nondimensional forms of the *mass balance equations* of the *TCE contaminant* the *water* and *air contaminants,* and the *linear momentum balance equation (Darcy's law)* for each $l$ *phase*

$$\begin{aligned}
&\mathrm{Sh}\frac{\partial}{\partial t'}\left(\phi S_l M_l X_{c_m l} + \rho_b \kappa_d M_l X_{c_m l}\delta_{lq}\right) = \nabla' \cdot \left(\mathrm{Pe}^{-1}\phi S_l \mathbf{D}'_{c_m l} M_l \nabla' X_{c_m l} - M_l X_{c_m l}\mathbf{V}'_l\right); \quad l \equiv n,q,g \\
&\mathrm{Sh}\frac{\partial}{\partial t'}\left(\phi S_l M_l X_{wl}\right) = \nabla' \cdot \left(\mathrm{Pe}^{-1}\phi S_l \mathbf{D}'_{wl} M_l \nabla' X_{wl} - M_l X_{wl}\mathbf{V}'_l\right); \quad l = q,g \\
&\mathrm{Sh}\frac{\partial}{\partial t'}\left(\phi S_g M_g X_{ag}\right) = \nabla' \cdot \left(\mathrm{Pe}^{-1}\phi S_g \mathbf{D}'_{ag} M_g \nabla' X_{ag} - M_g X_{ag}\mathbf{V}'_g\right) \\
&\mathbf{V}'_l = -F\,\mathrm{Re}\frac{k_{rl}}{\mu'_l}\mathbf{K}'\left(\mathrm{Eu}\nabla' P'_l + \frac{\rho'_l}{\mathrm{Fr}}\nabla' Z'\right); \quad \forall_l
\end{aligned} \tag{5}$$

In (5), $\phi$ denotes the porosity, $\rho_b[kg\,m^{-3}]$ denotes the solid matrix mass density, $\kappa_d[m^3\,kg^{-1}]$ denotes the sorption coefficient, $\delta_{ij}$ denotes the Kronecker delta function, $\mathbf{D}_{pl}$ denotes the dispersion tensor for the $p$ component in the $l$ phase, defined by

$$\phi S_l \mathbf{D}_{pl} = \left(\phi S_l D_{pl} + \alpha_{lT} \left|\mathbf{V}_l\right|\right)\delta_{ij} + \left(\alpha_{lL} + \alpha_{lT}\right)\frac{V_{li}V_{lj}}{\left|\mathbf{V}_l\right|}, \quad \forall l \qquad (6)$$

where $D_{pl}[m^2 s^{-1}]$ denotes the diffusion coefficient of $p$ in $l$, and $\alpha_{lT}, \alpha_{lL}[m]$ denote the transverse and longitudinal dispersivities, respectively, in $l$.

Note that, $q_p[mole\, m^{-3}\, s^{-1}]$, a source/sink mass flux of the $p$ component is not expressed in (5) for the sake of simplicity. By virtue of (5) we note that the phases and components temporal and spatial distribution is also governed by $\mathrm{Sh}$, $\mathrm{Pe}$, $(F\,\mathrm{Re}\,\mathrm{Eu})$ and $(F\,\mathrm{Re}\,\mathrm{Fr}^{-1})$ numbers, where $\mathrm{Sh} \equiv \left((L/T)/V_l\right)$ denotes the Strouhal number, $\mathrm{Pe} \equiv \left[\left(V/D_p\right)_l L\right]$ denotes the Peclet number, $\mathrm{Re} \equiv \left[(V/(\mu/\rho))_l L\right]$ denotes the Reynolds number, $\mathrm{Eu} \equiv \left(P/(\rho V^2)\right)_l$ denotes the Euler number, $\mathrm{Fr} \equiv \left((V_l^2/L)g\right)$ denotes the Froud number and $F \equiv \left(K/L^2\right)$. If the scalar numbers $\mathrm{Sh}$, $\mathrm{Pe}$, $(F\,\mathrm{Re}\,\mathrm{Eu})$ and $(F\,\mathrm{Re}\,\mathrm{Fr}^{-1})$ configured at one spatial scale are made equal to their counterpart numbers obtained at a domain ,say, greater in its spatial scale, one can expect similar flow and transport characteristics in both scales.

Next we will provide examples describing migration patterns resulting from the coupled affect between phases and components when subject to a variety of influences.

## Implementations and conclusions

### Two Dimensional simulations

To study the effect of the different boundary and initial conditions, operation scenarios and geological formation on the migration of the TCE plume, simulations using synthetic data were considered for a 2-D cross section configuration. The layout of the modelling domain.is depicted in Fig. 1. Extending from the domain left boundary are three clay layers with assumed absolute permeability of $K=10^{-14}$ m$^2$ each. The absolute permeability of sand was prescribed to be $K=3\ 10^{-11}$ m$^2$.

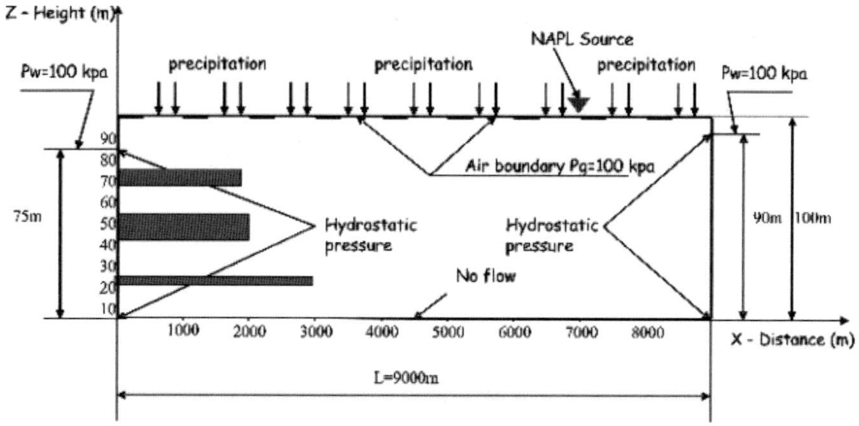


**Fig. 1.** Computational domain for the two-dimensional, layered, problem, x=0 corresponds to distance of 1500 m from seashore.

The annual replenishment was set to 100 mm for 100 (winter) days. Segments (0% to 80% coverage of the top boundary) of the soil surface were assigned to be impervious to replenishment. Gas was allowed to penetrate through the entire soil surface. Note that components (e.g. TCE) could be displaced outside the domain, as the impervious segments were not assigned a no-flux condition for the fluid phases. The initial aqueous and gaseous regime was determined by transient simulation evolving from a non-stressed aquifer (zero saturation for the non-aqueous phase) and a water table situated at the soil surface. Stabilization of the system took a 10 years period. Forecast was for the next 50 years. NAPL was deployed within one cell of 50 m length ($6950 \leq x \leq 7000$ m), situated on the surface (Z=100 m). Injection into the soil was constant, over a period of 50 years, with a flux intensity of $q_R$ = 0.001 $m^3/m^2$/day. To verify the influence of surface coverage we performed five simulations each with a different percentage (0, 20, 40, 60, and 80%) of surface cover.

Fig. 2 depicts the distribution of phases saturation and TCE mole fraction in the gaseous and aqueous phases for 80%, surface coverage. We note the generation of a thin vertical column of NAPL under the spill. Due to gravity, the NAPL front reached the bottom of the domain in about 10 years. During 40 years it moved downstream (flow direction of the aqueous phase) along the impermeable bottom to a distance of 500 m (460 m for 0% coverage) This displacement was under the influence of the NAPL, saturation dependent, gradient difference between the downstream aqueous pressure and the capillary NAPL pressure. The upstream displacement along the bottom was 40 m (70 m for 0 % coverage).

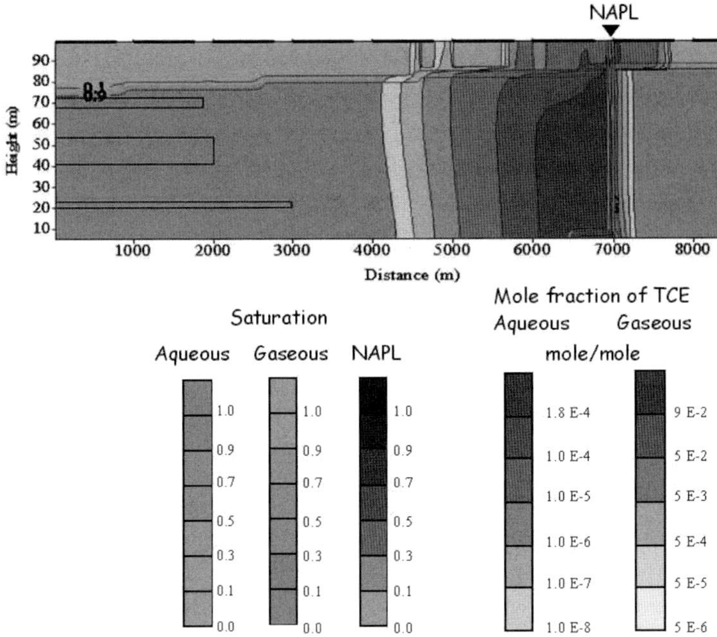


**Fig. 2.** Saturation and TCE mole fraction in the aqueous and gaseous phases, for the case of 100 mm replenishment over 80% surface coverage in a cyclic pattern, during 100 days per year for 50 years

After 50 years most of the TCE contaminant mass was accumulated in the non-aqueous phase. In the aqueous phase, the concentration profile along the vertical direction became practically uniform. The mole fraction of the TCE changes dramatically crossing along the Saturated–Unsaturated Interface Region. By virtue of Henry's law, TCE has a continuous mole fraction distribution within the aqueous and the gaseous phases yet it is of a lower value in the aqueous phase. Downstream displacement of the maximum TCE concentration in the aqueous phase (1.8 $10^{-4}$ mole/mole) was 600 m for the uncovered surface (i.e. 0% coverage), and 900 m for the 80% covered surface. Maximum TCE migration was about 2700-2800 m in groundwater and 2600 m in the unsaturated zone. Upstream TCE displacement in groundwater is similar for both possibilities (0% and 80% coverage). However, in the unsaturated zone the TCE migration distance is to a greater extent for the 80% coverage. Transport of TCE is governed by the phases flow regime. The constant hydrostatic pressure conditions at the right and left boundaries control the hydraulic gradient. However within the domain, the steepness of this gradient is affected by the surface coverage conditions.

The velocity vectors of the aqueous phase are delineated in Fig. 3.A.

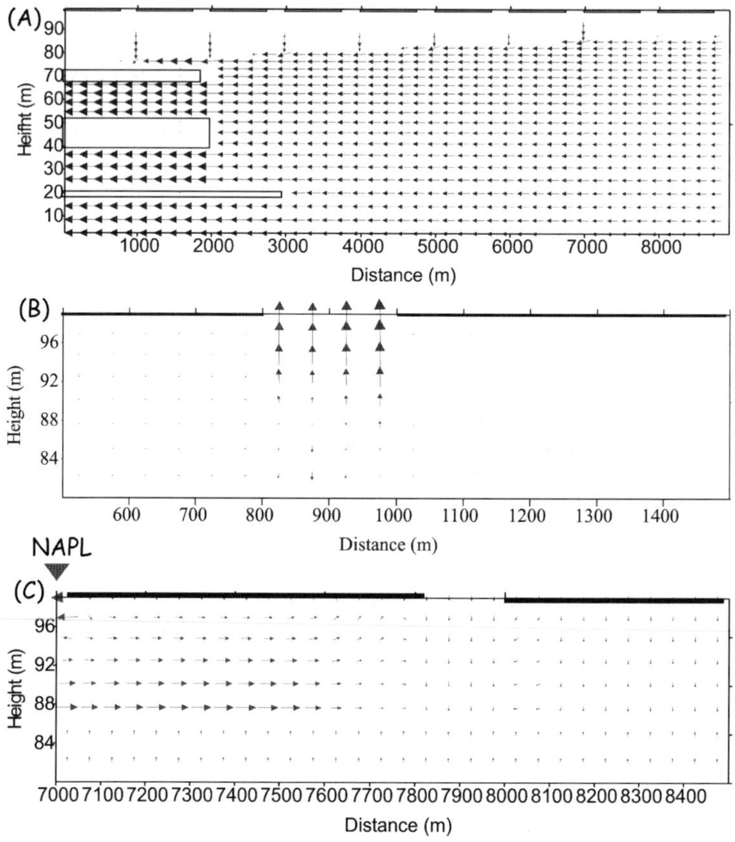


**Fig. 3.** 100 mm replenishment over 80% surface coverage in a cyclic pattern, during 100 days per year for 50 years. (A) Aqueous velocity pattern. (B) Gaseous flow through an opening during a rainy season. (C) Gaseous flow near the NAPL spill

The typical gaseous flow pattern through the surface uncovered areas is depicted in the snapshot presented in Fig. 3.B. Through these openings infiltration of seasonal rain causes an oscillatory displacement of the aqueous and gaseous phases. The gaseous phase is displaced away from the NAPL phase; its specific flux near the spill is smaller for the uncovered surface case than that of the 80% coverage. At the center of the spill (i.e. along the NAPL column) the NAPL gravity shift causes a vertical gaseous phase velocity also influenced by the gaseous pressure gradient (equal to the gradient of NAPL and capillary pressures). As TCE concentration decreases away from the NAPL column and since the phases densities are function of TCE concentration, the density gaseous gradient yields its horizontal specific flux (the blowup described in Fig. 3.C) and its vertical outward shift results from the vanishing capillary pressure. Far away from the NAPL column the gaseous velocity is associated with gravity and there are no density driven effects (density is constant as TCE vanishes) and it is diminished at the gaseous-aqueous interface, as its permeability there is zero.

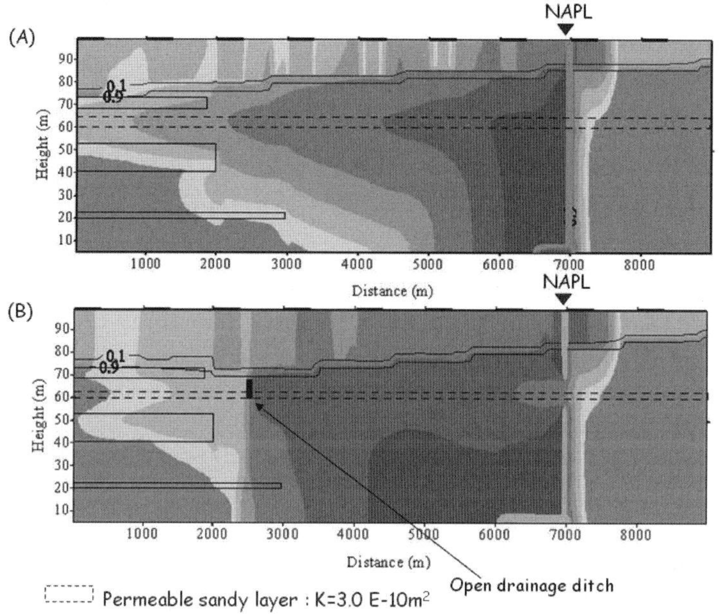


**Fig. 4.** 100 mm replenishment during 100 days per year over 60% surface coverage in a cyclic pattern, with a preferential flow path. (A) TCE plumes after 50 years. (B) Remediation, due to drainage through an open ditch, for additional 30 years.

To investigate the roll of a preferential flow path, we added a layer with absolute permeability (K=3 $10^{-10}$ $m^2$) higher than that of the domain. Conditions and replenishment (100 mm per 100 days of each year, over 60% coverage in a cyclic pattern of the soil surface), were as before. Fig. 4.A depicts the extent of TCE migration, centered at the preferential flow path, after a 50 years period.

After the 50 years period of contamination we placed within the preferential flow path an open ditch; Fig. 4.B. describes the extent of the remediation due to the drainage through the ditch. The aqueous flow is augmented at the surroundings of the low permeable layers (Fig. 3.A). This flow pattern forces the amplification of TCE advection, its penetration into the layers due to the dispersion flux and a sideway concentration kink due to the eddy action adjacent to the layers corners (Fig. 4.B).

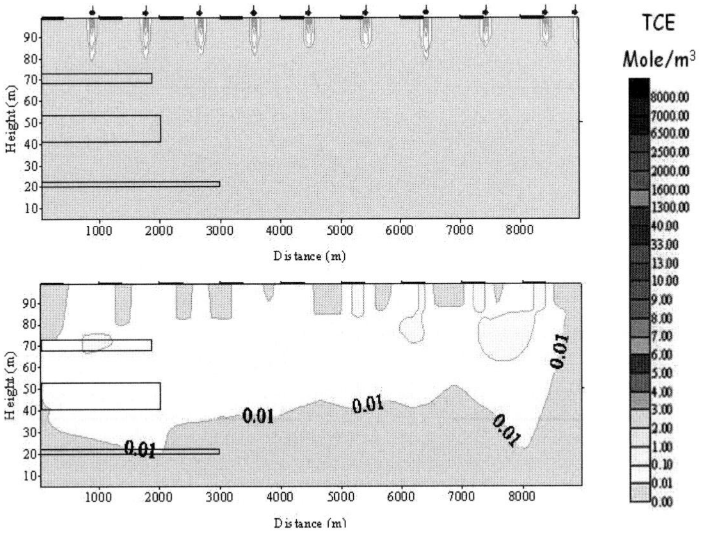


**Fig. 5.** 100 mm replenishment during 100 days per year, over 40 % surface coverage in a cyclic pattern. (A) Distribution of TCE due to spills lasting 100 days. (B) Distribution of TCE after 50 years.

Wc now dcployed spills for a short duration (100 days), to mimic the notion that there are NAPL residuals in the soil matrix. This can also rcpresent an accidental spill from different sources. The NAPL sources were spread over 10 different locations, uniformly distributed along the soil surface. The previous injection intensity was divided over the 10 locations. Fig. 5.A describes the deployment of the NAPL source for a period of 100 days. Each injection occupied one grid element (50 m), initial and boundary conditions were as for the previous examples, the intensity and distribution of replenishment were as prescribed for the previous example. During the first 100 days (Fig. 5.A), the NAPL occupied an extent of 1 to 3 grid elements (2 m to 6.5 m) under the injection point, which practically, because of the small mass amount, enabled us to consider TCE concentration in [mass/volume] units. Saturation of NAPL was less then 0.08, therefore as the residual NAPL saturation is 0.05 (for which its relative permeability is zero) we conclude that residual NAPL remains adhered to the soil matrix in the unsaturated zone. It is evident (Fig. 5.B) that in contrast to the high TCE concentration during the first 100 days, after 10 years TCE was significantly dissolved.

We note (Fig. 5.B) that the neighboring TCE spills joined into a continuous plume (concentration higher than 0.01 mole/m$^3$), disabling the detection (practically after 10 years) of the individual spill sources. The pronounced aqueous specific flux in between the layers (Fig. 3.A) yields an advective flushing of TCE surrounding the layers; yet (Fig. 5.B) due to the intense dispersion flux at that vicinity, a residual of higher TCE concentration was trapped inside one of the low permeability layers.

## Three Dimensional simulations of the study area

The 5 available hydrogeological cross-sections were schematized delineating three main clay layers that divide the main aquifer into 4 sub-aquifers. Interpolation between the cross sections, on the basis of 10 vertical profiles for each cross section enabled an approximate 3-D stratigraphic model of lithological properties and cross-section images anywhere within the investigated domain. A 3-D finite-elements non-uniform mesh (57, 58 and 20 nodes in the x, y and z-directions, respectively), refined at the location of the NAPL source and with vertical variation between 4 to 8 m (depending on the aquifer thickness) was constructed on the basis of the stratigraphic model (Fig. 6).

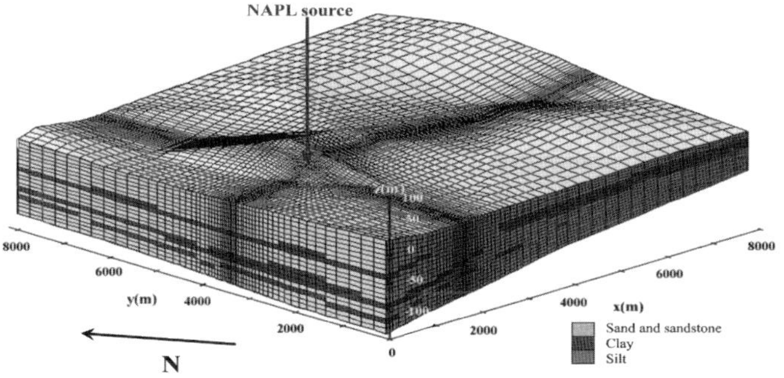


**Fig. 6.** The modeling domain with the Finite-Element mesh (66,120 nodes)

Available data concerning the location of 106 wells, their pumping rate during the years 1924 to 2000, was gathered into 20 cluster wells (CW) using a partitioning method (Kaufman and Rousseeuw 1990). The CW distribution and the boundary conditions in the modeling domain are described in Fig. 7. Based on results of sensitivity analysis, we chose the permeability of sand and sandstone as the parameter which was found to be mostly affecting groundwater levels and calibrated the flow model. The comparison between observed and simulated groundwater levels for the year 1980, is presented in Fig. 8. In reality, the actual total amount of NAPL, the size of the discharge area, the discharge intensity and discharge continuity are not known. Taking this into account, we considered 3 different rates of NAPL discharge: 1, 0.2 and 0.02 mm/day, and simulated the multiphase flow and TCE transport during the period 1950 to 2000. In our simulations we assumed that NAPL phase consists of TCE only. For all three cases the aerial extent of the TCE plumes were found to be mimicking the characteristics of the observed TCE plume, but with different distribution patterns. The distribution of TCE concentration in the unsaturated and saturated zones expressed in mole of TCE per mole of aqueous phase is delineated in Fig. 9. Because of the high density of the NAPL phase, its migration through the unsaturated and saturated zones is gravity-driven, and its general displacement is influenced by the hydraulic conductivity, and slope of the different layers. In the unsaturated zone, the NAPL front migrates mainly downwards with an average velocity that decreases upon entering the saturated zone and further attenuates when it reaches the clay layer. Dissolution and volatilization of the NAPL takes place during its movement causing the TCE to be transferred to the aqueous and gaseous phases. Lateral TCE transport through the unsaturated zone occurs mainly in the gaseous phase as the NAPL source is situated within the area covered by an upper impervious boundary (e.g., pavement; Fig. 7). To the west, east and south directions, the velocity of the gaseous phase is higher than groundwater velocity; therefore, secondary contamination of groundwater by TCE dissolving from the gaseous phase is evident.

To the north, groundwater velocity exceeds the velocity of the gaseous phase; therefore, secondary contamination of the unsaturated zone occurs by volatilization of TCE from groundwater far away from the NAPL source (Fig. 8). The transport of TCE in the saturated zone is dominated by advection of groundwater flow, which is directed to the north-east, towards a local groundwater depression. However, the maximum migration of the TCE plume to the south was predicted in the upper part of the aquifer as a result of secondary groundwater contamination. Less intensive NAPL discharge yields less penetration depth, yet migration of TCE in the gaseous phase in the unsaturated zone is similar to the harshest discharge. Additional scenario was simulated for a case when the soil surface was not covered.

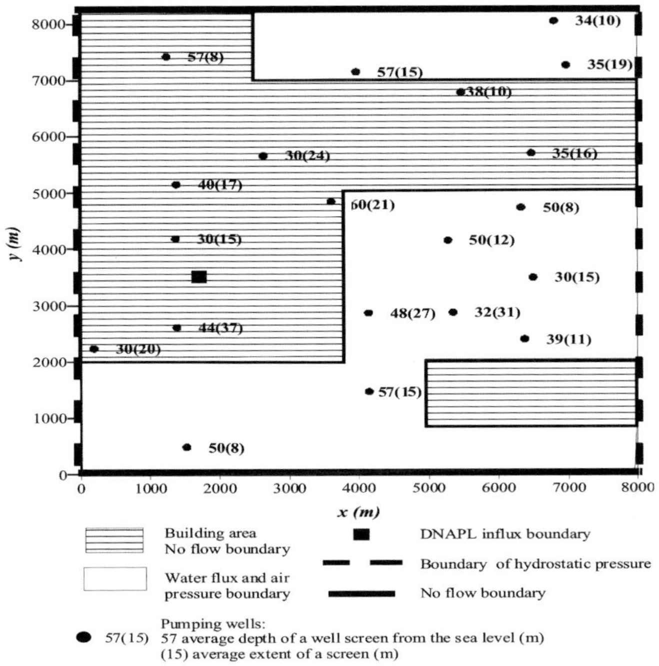


**Fig. 7.** Schematic representation of boundary conditions in the modeling domain

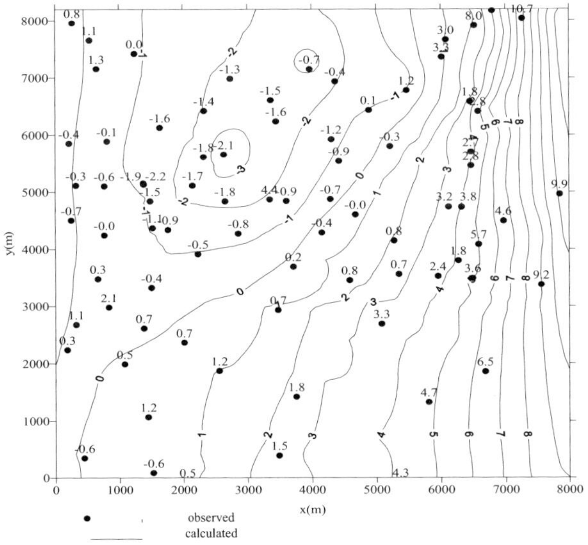


**Fig. 8.** Comparison between observed (numbers above observation points) and simulated (lines) groundwater levels for the year 1980

Obtained results indicate a significant decrease in the lateral spread of TCE through the unsaturated zone as compared to the case where part of the surface was covered. This is accounted by the fact that the gaseous phase can exit to the atmosphere which also decreases the migration of TCE in groundwater.

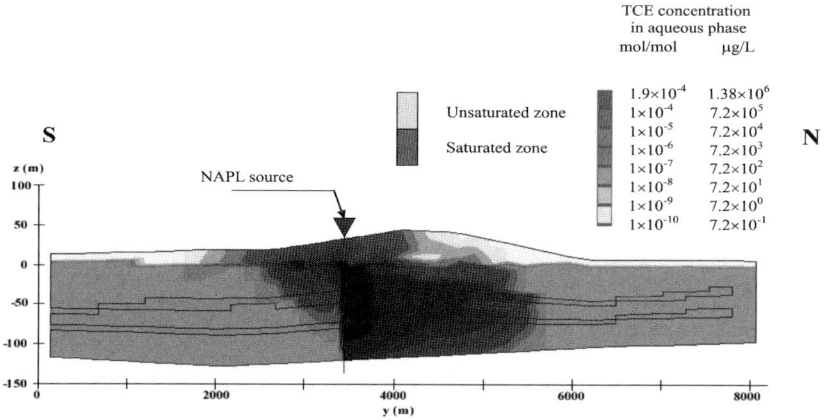


**Fig. 9.** TCE plumes through a South-North cross-section for the year 2000, subject to the harshest, simulated, NAPL discharge (1 mm/day)

## References

Forsyth PA, Unger AJA, Sudicky EA (1998) Nonlinear iteration methods for non-equilibrium multiphase subsurface flow. Adv. Water Resour., on of LNAPL-contaminated aquifers, J Cont Hydrology, 31, pp 57-81

Kaufman L, Rousseeuw PJ (1990) Finding Groups in Data: An Introduction to Cluster Analysis. Wiley, New York

Ronen D, Scher H, Blunt M (2000) Field observations of a capillary fringe before and after a rainy season. Journal of Contaminant Hydrology, Vol. 44, pp 103-118

Slough KJ, Suddicky EA, Forsyth PA (1996) Grid refinement for modeling multi-phase flow in discretely fractured porous media. Adv Water Resour, 23, pp 261-269

Slough KJ, Suddicky EA, Forsyth PA (1999) Numerical simulation of multiphase flow and phase partitioning in discretely fractured geologic media. J Cont Hydrology, 40, pp 107-136

Unger AJA, Forsyth PA, Sudicky EA (1996) Variable spatial and temporal weighting schemes for use in multi-phase compositional problems", Adv Water Resour, Vol. 19, No. 1, pp 1-27

Unger AJA, Forsyth PA, Sudicky EA (1998) Influence of alternative dissolution models and subsurface heterogeneity on DNAPL disappearance times. J Cont Hydrology, 30, pp 217-242

# Solute contaminant transport in variably saturated dual-porosity/dual permeability chalk: field tracer experiments and modelling

S. Brouyère

Hydrogeology and Environmental Geology, Dept of Georesources, Geotechnologies and Building Materials (GeomaC), University of Liège - Belgium
Building B52/3, 4000 Sart Tilman, Belgium
Tel: +32.43662377 - Fax: +32.43669520 – Email: Serge.Brouyere@ulg.ac.bet

## Abstract

A detailed field investigation was performed for studying groundwater recharge processes and solute downward migration mechanisms prevailing in the unsaturated zone overlying a chalk aquifer in Belgium. Various laboratory and field experiments were performed, among others tracer experiments in the unsaturated zone. From the experiments performed in the variably saturated chalk, it appears that the migration and retardation of solutes is strongly influenced by recharge conditions. Under intense injection conditions, solutes migrated at high speed along the partially saturated fissures, reaching the saturated zone in a few hours. At the same time, they were temporarily retarded in the almost immobile water located in the chalk matrix. Under normal recharge conditions, fissures were inactive and solutes migrated slowly through the chalk matrix, taking about one year to reach the water table. The tracer experiments performed in the chalk were modelled using the finite element flow and transport simulator SUFT3D. The mathematical representation of the unsaturated properties of the fissured chalk is based on a new modified van Genuchten – Mualem relationship that takes into account the dual-porosity characteristics of the rock. For modelling the transport of the tracers, an adapted first-order transfer, dual-porosity model is used, based on a dynamic partitioning of water between effective and immobile water porosity according to the water saturation of the chalk. Conclusions are drawn in terms of recharge mechanisms and vulnerability of the chalk to pollutions occurring at the land surface.

## Keywords

Dual porosity; dual permeability; chalk; unsaturated zone; transport; tracer

## Introduction

The Hesbaye aquifer is located in the Senonian chalk formations of the Geer basin in the eastern part of Belgium. Since 1960, nitrate concentrations have risen annually at a rate ranging from 0.1 mg/l (as NO3-) in the semi-confined to 1 mg/l in the unconfined part of the aquifer. Presently, the mean concentration is around 35 mg/l. To have a better knowledge and understanding of groundwater recharge mechanisms and transport processes in the unsaturated zone of the Hesbaye aquifer, an experimental study was performed, funded by the Walloon Ministry of Natural Resources and Environment. The experimental site is located at Bovenistier in the Hesbaye region. From top to bottom, the substratum is made of quaternary loess of variable thickness (up to 20 m), locally several meters of tertiary sand deposits, a layer of flint conglomerates (maximum 10 m), up to 70 m of Senonian chalk and several meters of smectite clay. Six boreholes were drilled. A central well (PC) was drilled down to the aquifer bottom (50 m). An observation well (Pz CS) was drilled and screened in the saturated zone (34 m). Four subsequent wells were drilled and screened in the unsaturated zone: Pz CNS in the unsaturated chalk (19.2 m), Pz CGL in the flint conglomerate (15.8 m) and Pz LB (9.4 m) and LS (5.8 m) in the loess formations. For each borehole, core drilling was first performed to take undisturbed samples used for performing laboratory experiments such as hydraulic conductivity measurements, texture and density analyses, retention curves measurements and pesticides and nitrate analyses. In the field, well logging (neutron, gamma-gamma and gamma ray) was performed in each well. Pumping tests were performed in the boreholes screened in the saturated zone and infiltration tests in the boreholes screened in the unsaturated zone. Tracer experiments were performed in both the saturated and the unsaturated zone. Details about the experiments can be found in Brouyère et al. (2004).

## Tracer experiments performed in the variably saturated chalk

Two tracer tests were performed in the saturated zone, under radially converging flow conditions, between Pz CS and the central well (tracing distance: 6 m). For the first injection the pumping rate at the central well was 1.2 m3/h, for the second, 6.0 m3/h. The breakthrough curves highlighted the very fast transit of tracers in the saturated zone, mostly along fissures, with a strong asymmetry and tail that reflected tracer dispersion and retardation in the immobile water located in the chalk matrix (results not shown). Tracers were also injected in Pz CNS screened in the unsaturated chalk. During the two years of monitoring, pumping was maintained almost permanently at the central well (3.2 to 6 m3/h). During the period of the experiments, the thickness of unsaturated chalk was progressively reduced from 11 to 3 m because of a rise of groundwater levels. A first injection (potassium chloride KCl) was performed under artificial recharge conditions created by adding water at a constant rate (~0.1 m3/h). This led to enhanced hydraulic gradient between the injection point and the aquifer and to a higher degree of saturation in the unsaturated chalk. A second tracer injection (potassium iodide KI) was performed without addition of water. This configuration better reflects actual seepage conditions in the unsaturated chalk. In the first case, the tracer was detected 5h 15min after injection, the maximum concentration being reached 11h 30min after injection (Fig. 1a). In the second case, transfer times were very different: the tracer was detected only after 340 days while, after 700 days (end of the monitoring campaign), the peak was still not reached even if concentrations seemed to stabilize (Fig. 1b).

To explain the strong contrast in terms of travel times (about three orders of magnitude), the dual porosity – dual permeability of the chalk has to be invoked. The chalk matrix is characterised by a high porosity and a low hydraulic conductivity at saturation. At Bovenistier, laboratory experiments performed on chalk matrix samples gave a matrix porosity ranging from 38 to 51 %, and a saturated hydraulic conductivity of 1.3×10-8 m/s. It is made up of a large number of pores having dimensions of the order of m, in which capillary forces are very high. As a direct consequence, the□1 matrix remains close to saturation and water is almost immobile. On the contrary, the porosity associated with the fissures is very low but their contribution to the saturated hydraulic conductivity of the rock is predominant. Here, pumping tests and infiltration tests show similar orders of magnitudes for the hydraulic conductivity of the fissured chalk ($10^{5}$-$10^{-4}$ m/s). Fissures are characterized by larger openings and capillary forces are not so developed. Under natural recharge conditions, they are most often empty and inactive in the unsaturated zone. The average velocity of downward migration of water and solutes in the unsaturated chalk is very low, governed by the low hydraulic conductivity and the large porosity of the matrix. During periods of intense water recharge, it may occur that the matrix becomes fully saturated. The excess of water is free to infiltrate through the fissure network with a quick increase of the chalk hydraulic conductivity. Water then flows mostly through the fissure network, the role of the matrix being limited to the storage and release of water and solutes. Under such conditions, any contaminant present in the infiltrating water is able to reach very quickly the water table by fast downward migration through the fissure network.

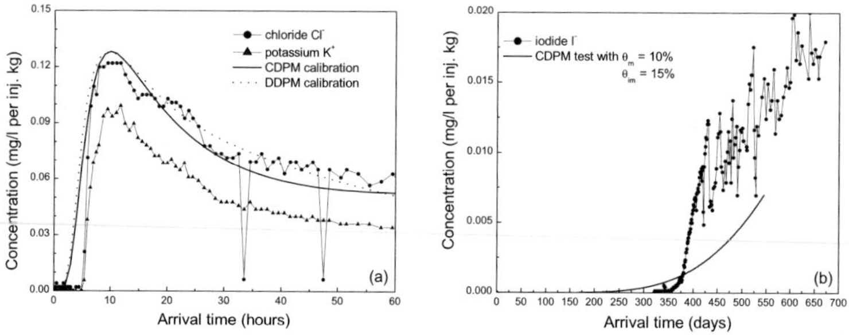


**Fig. 1.** Breakthrough curves obtained for the tracer experiments performed in the unsaturated chalk and modelled using the SUFT3D (see section 4 for details).

## Mathematical model

To transcribe mathematically the conceptual model presented here above, Brouyère (2001) developed an approach that relies on two main concepts. The unsaturated properties of the chalk are modelled using a modified van Genuchten relationship that accounts for the bimodal porosity of the rock. The retention curve of the chalk is modelled using the combination of two van Genuchten relationships, one defined for the matrix, one for the fissures.

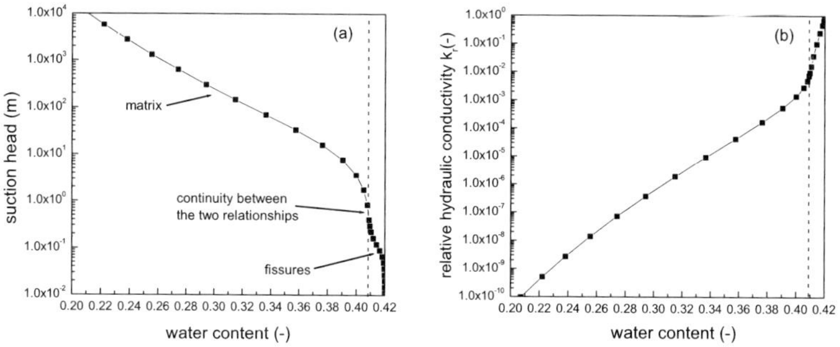


**Fig. 2.** Retention curve (a) and Hydraulic conductivity curve (b) developed for the fissured chalk

This model differs from that of Ross and Smettem (1993) or Durner (1994) by the fact that it forces a continuously derivable transition between the relations that describe the matrix and the fissure component respectively (Fig. 2a). Thanks to that, it is possible to derive analytically the hydraulic conductivity curve from the retention curve for the whole range of water contents (Fig. 2b), using the model of Mualem (1976). The hydraulic conductivity curve shows the expected evolution with water content or suction head. For a small suction applied, fissures desaturate and the hydraulic conductivity of the chalk rock drops by several orders of magnitude.

The dual-porosity first order transfer model is a very convenient model to compute contaminant transport and retardation in fissured rocks (e.g. Brouyère et al. 2000). This model introduces two parameters in the calibration process: the dual porosity $\theta_{im}$ (no units) and a first order transfer coefficient $\alpha$ ($T^{-1}$). In its original form, the porosity associated with the immobile water is assumed constant. However, when the rock mass desaturates, its hydraulic conductivity is reduced. Water located in the dual porosity and initially considered as immobile becomes the only vector of mobility of contaminants. In transient variably saturated groundwater flow conditions, the model has to be adapted by distributing dynamically water between the mobile and the immobile phases, depending of the saturation degree. Zurmülh and Durner (1996) suggest considering a constant ratio between the immobile water porosity and the total water content $c_{part} = \theta_{im}/\theta$ or distributing the water according to the ratio of intrinsic hydraulic conductivity values associated with the immobile water and the total water content $cpart = kr(\theta_{im})/kr(\theta)$.

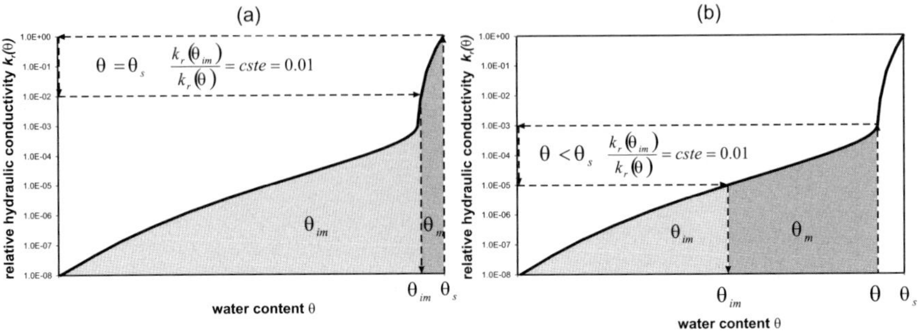


**Fig. 3.** Dynamic partitioning of the water content between the mobile and the immobile water according to the saturation degree of the chalk rock: (a) matrix and fissures are saturated with water, (b) fissures are empty and the matrix is partially saturated.

Figure 3 illustrates this concept considering a partitioning coefficient of $c_{part} = 0.01$. At saturation, the effective porosity is small, associated with the fissures, and the dual-porosity is important, associated with the matrix (Fig. 3a). When fissures desaturate, the effective porosity becomes larger, associated with a part of the water in the matrix, and the dual porosity is strongly reduced (Fig. 3b). This model was implemented in the SUFT3D finite element code (Brouyère 2001).

## Modelling results

The two tracer experiments performed in the saturated chalk, between Pz CS and the central well, were used to calibrate hydrodispersive parameters in the saturated zone (Table 1). The calibrated effective porosity is low, typical of the porosity associated to fissures in the chalk. The first-order transfer coefficient is in a good agreement with values found for the fissured chalk in the Hesbaye aquifer. The immobile water porosity is relatively low. However, the tracers were injected very close from the recovery well and it is likely that a part of the immobile water did not participate to the retardation process. The tracer experiment performed in the unsaturated chalk under intense injection conditions was modelled assuming steady state groundwater flow conditions, as modified by the injection of water in Pz CNS. First, the "classical" dual-porosity model (further called the CDPM) was considered, assuming that effective porosity and immobile water porosity are constant values independent of the chalk saturation degree. Second, the dynamic partitioning dual-porosity model (further called the DDPM) was considered. In this case, one does not know exactly the magnitudes of effective porosity and immobile water porosity, the only information being the partitioning coefficient considered as a fitting parameter. Figure 1 shows the chloride breakthrough curve measured in the field together with the best fit obtained (trial-and-error) considering the two approaches considered. Fitted parameters obtained using the CDPM approach are close to those obtained in the saturated zone. This indicates a fast, preferential transit of tracers along partially or fully saturated fissures with physical retardation in the immobile water located in the matrix. With the DDPM approach, the relatively high partitioning coefficient indicates that the effective porosity is small and the immobile water porosity large. The calibrated first order transfer coefficient is smaller than for the CDPM approach because the saturation degree varies spatially. Due to that, there are zones in the unsaturated chalk where the hydraulic conductivity is smaller, which contribute to retarding the tracer.

**Table 1.** Fitted transport parameters for the variably saturated chalk using the tracer test results.

| Injection well | Experimental conditions | Modelling approach | $\theta_m$ (-) | $\theta_{im}$ (-) | $\alpha_L$ (m) | $\alpha$ (s-1) |
|---|---|---|---|---|---|---|
| Pz CS | Saturated zone | CDPM | 0.004 | 0.05 | 1.0 | 2.0×10-7 |
| PsCNS | Intense recharge | CDPM | 0.01 | 0.08 | 1.0 | 2.3×10-7 |
| | | DDPM | cpart = 0.02 | | 1.0 | 9.0×10-8 |
| | Natural infiltration | CDPM | 0.10 | 0.15 | 1.0 | 2.3×10-7 |
| | | DDPM | no tracer arrival observed during the simulated period | | | |

The second tracer experiment is more representative of natural downward solute contaminants migration in the unsaturated chalk. For several reasons, simplifying assumptions were considered. First, the transient character of the recharge was neglected because no data were available on the evolution of water contents in the unsaturated zone. In addition, steady state groundwater flow conditions were assumed, considering the low groundwater levels prevailing at the beginning of the tracer experiments. The objectives of the transport simulations were thus restricted to verify that the strong contrast observed in terms of travel times across the unsaturated chalk could be reproduced and to estimate the effective porosity associated to the chalk in natural infiltration conditions. Using the DDPM approach with the set of hydrodispersive parameters calibrated based on the first tracer experiment or with an effective porosity of 35% (all the water located in the chalk is mobile) did not give any tracer arrival in the period corresponding to the sampling campaign (2 years). Considering an effective porosity of 10% and an immobile water porosity of 15% gave a tracer first arrival in accordance with the observed one (after almost one year). The value obtained for the effective porosity (10%) is small to be representative of the matrix porosity. However, if one considers that the travel distance (+/- 10 m) in the model was almost 3 times the actual distance covered by the tracer (+/- 3 m), a rough estimation of 30% can be assessed for the effective porosity, in better accordance with what is expected for the chalk matrix porosity. From the analysis of these simplified simulations, it can be concluded that in natural infiltration conditions, solute contaminant downward migration occurs across the chalk matrix, as proposed in the conceptual model.

## Conclusions

Detailed field investigations performed at Bovenistier in the Hesbaye region have provided a very complete dataset for understanding and quantifying water recharge and solute leaching processes across the unsaturated zone overlying the Hesbaye fissured chalk aquifer. In 'normal' infiltration conditions, the recharge is not sufficient to keep the fissure network at saturation. Water infiltrates slowly downward across the matrix and solute transport is influenced by the low hydraulic conductivity and the large porosity of the matrix. As soon as the net infiltration rate is higher than the saturated hydraulic conductivity of the matrix, the excess water is free to infiltrate rapidly through the fissure network. Compared to that, groundwater located in the matrix can be assumed as immobile, contributing only to storage and release of water and solutes, just like in the saturated zone. Based on the results, it can be concluded that the vulnerability of the chalk aquifer to contaminations occurring at the land surface is strongly dependent on the existence of a protective cover. Due to that, the recharge rate at the top of the unsaturated chalk is attenuated and contaminants migrate slowly down through the matrix, enhancing the possibilities for natural attenuation. If the chalk was outcropping, any contaminant present in the infiltrating water would be able to reach very quickly the water table. This conceptual and mathematical model could be used to study the impact of groundwater table fluctuations on the evolution of contaminant concentrations such as nitrates, for the estimation of groundwater quality trends for which this factor has to be considered or as a mathematical translation of recent researches, such as those of Price et al. (2000) or Tokunaga et al. (2000).

## References

Brouyère S, Dassargues A, Therrien R, Sudicky E (2000) Modelling of dual porosity media: comparison of different techniques and evaluation on the impact on plume transport simulations. IAHS Publication n°265, ed. Stauffer F, Kinzelbach W, Kovar K & Hoehn E, IAHS Press, Wallingford, Oxfordshire, UK, pp 22-27

Brouyère S (2001) Etude et modélisation du transport et du piégeage des solutés en milieu souterrain variablement saturé (study and modelling of transport and retardation of solutes in variably saturated media), in French. PhD Thesis, University of Liège, Faculty of Applied Sciences, pp 572

Brouyère S, Dassargues A, Hallet V (1994) Migration of contaminants through the unsaturated zone overlying the Hesbaye chalky aquifer in Belgium: a field investigation. J Contam Hydrol, in press

Durner W (1994) Hydraulic conductivity estimation for soils with heterogeneous pore structure. Water Resour Res, 30(2):211-223

Mualem Y (1976) A new model for predicting the hydraulic conductivity of unsaturated porous media. Water Resour Res, 12(3):513-522

Price M, Low RG, McCann C (2000) Mechanisms of water storage and flow in the unsaturated zone of the Chalk aquifer. J Hydrol, 233:51-74

Ross PJ, Smetten RJ (1993) Describing soil hydraulic properties with sums of simple functions. Soil Sci Soc Am J, 57:26-29

Tokunaga KT, Wan J, Sutton SR (2000) Transient film flow on rough fracture surfaces. Water Resour Res, 36(7):1737-1746

van Genuchten MTh (1980) A closed-form equation for predicting the hydraulic conductivity of unsaturated soils. Soil Sci Soc Am J, 44:892-898

Zurmühl T, Durner W (1996) Modeling transient water and solute transport in biporous soil. Water Resour Res, 32(4):819-829

# Integration of pedotransfer functions and topographical data to obtain soil hydraulic properties at catchment scale

M. Palladino, N. Romano, A. Santini

Department of Agricultural Engineering, University of Naples Federico II, Portici (Naples) - Italy

## Abstract

The emerging importance of large scale distributed-process modeling has generated a pressing need for detailed information on spatial distribution of input data, especially the parameters that describe the hydrologic behavior of soils. Traditional methods to determine soil hydraulic parameters at large scales are time-consuming and very costly. To carry out a soil hydraulic characterization in a cost-effective way, pedotransfer functions (PTFs) are profitable tools to estimate soil hydraulic parameters from easily measurable or already available soil data. Spatial distribution of soil properties can be explained, to a certain extent, in the light of the variability of landscape attributes. Digital terrain analysis can thus provide a quantitative basis for deriving topographic attributes and relating them to soil variables in order to improve the prediction of key soil hydraulic parameters. Topographical data have been used to improve the prediction of soil water retention by PTFs, whereas statistical indicators of goodness-of-fit helped in evaluating the effectiveness of the proposed procedure.

## Keywords

Soil hydraulic properties; pedotransfer function; Digital Elevation Model; terrain analysis; spatial variability

## Introduction

The use of distributed hydrologic modeling has been proposed to deal with environmental and land-use planning problems in a comprehensive way (Refsgaard 1997). However, distributed modeling approach has generated an urgent need for detailed information on spatial input data. There is the awareness that reliable results from distributed hydrologic models rely upon the good assessment of the parameters that describe the hydraulic behavior of soil. In fact, soil hydraulic properties and their inherent spatial variability exert a significant influence on the exchange of water fluxes between the different parts on the system. Moreover, applications of large-scale hydrologic models require not only a careful analysis of spatial correlations among soil parameters, but also an assessment of equivalent grid-scale parameters, which can be defined as areally averaged values over the selected numerical grid square. To simulate water flow in soil, distributed models usually employ the Richards equation:

$$C(h)\frac{\partial h}{\partial t} = \frac{\partial}{\partial z}\left[K(\theta)\left(\frac{\partial h}{\partial z} - 1\right)\right] \qquad (1)$$

In this equation, z is soil depth taken positive downward, t is time, $\theta$ is volumetric water content, h is soil-water pressure head, K is the unsaturated soil hydraulic conductivity, and $C(h)=d\theta/dh$. Model parameters, the so-called unsaturated soil hydraulic properties, are the water retention function $\theta(h)$ and the hydraulic conductivity function $K(\theta)$, which synthesize the relationships between h, $\theta$, and K.

Unsaturated soil hydraulic properties can be determined by direct methods under either laboratory or field experimental conditions. It is well known that direct procedures are time-consuming, require complex measuring devices and skilled operators, and hence are very costly. To provide catchment models with soil hydraulic properties in a cost-effective way, pedotransfer functions (PTFs) are being used increasingly to estimate water transport parameters from easily measurable or already available soil data. Basically, a PTF relates soil hydraulic parameters to some physical-chemical properties of the soil, such as texture, oven-dry bulk density, organic carbon content, porosity, etc., that are measured at the same point scale of interest (Bouma 1989).

Several researchers have developed PTFs, based on site-specific field campaigns and comprehensive databases developed in Europe (Wösten et al. 1999) or in the USA (Nemes et al. 1999). Romano and Santini (1997) showed that spatial series of water retention points generated by PTFs led to shapes of the probability distribution that differ, in some cases even markedly, from those of the measured water retention data. On the other hand, they also found that a PTF could be able to catch efficiently the average water retention behavior of a certain area. Therefore, a limitation of published PTFs is that they should be employed with confidence only within restricted ranges of soil types and environmental conditions (Bastet et al. 1999). Furthermore, PTFs take only a limited number of explanatory static variables into account and, as a result, they will capture with difficulty the inherent dynamic behavior of a soil (Tietje and Hennings 1996).

Several studies have shown that the spatial horizontal distribution of soil properties is often not completely random, but a structure of spatial variability of such properties can be detected (Burrough 1993). A basic explanation of this occurrence considers that soils form as a result of climate, relief, parent materials, weathering, organisms, and time. Each of these soil-forming factors shows a peculiar variability in the landscape and acts at various spatial-temporal scales producing different soil types and complex soil patterns. Among the various soil-forming factors, topography exerts a significant control on hydrological processes active in the landscape and can explain a good amount of spatial variability at this scale (Famiglietti et al. 1998). Topographical features of a certain area can be quantified by the mean of terrain attributes. Primary terrain attributes can be calculated employing digital terrain analysis from a Digital Elevation Model (DEM), and include variables such as elevation, slope gradient, plan and profile curvature, flow path lengths, and specific catchment area. With combinations of primary attributes we can obtain secondary terrain attributes, that characterize the spatial patterns of specific processes occurring in the landscape (Moore et al. 1991). These topographic indices provide a knowledge-based approach to specific soil management and can be embedded within the data analysis module of a geographical information system (GIS). As many GISs are based on a raster structure, grid-based methods of terrain analysis can easily provide primary geographic data for GIS applications.

The focus of the present paper is on using terrain attributes as auxiliary variables to determine soil water retention and show that this approach can effectively improve the average description of soil hydraulic behavior over a certain area. Moreover, developing such a procedure can also give useful directions for upscaling more efficiently soil hydraulic parameters from the measurement scale to the numerical scale.

## Methods

### Pedotransfer functions

Several PTFs were used to evaluate the proposed procedure. For the sake of brevity, we report here only results from the pedotransfer developed by Vereecken (Vereecken et al. 1989), which will be referred to hereinafter as PTF-VER. Parameter equations which define this PTF are not illustrated here, but readers are directed to the cited publication for details.

### Terrain attributes

In this study, calculations of terrain attributes were carried out by digital terrain analysis using the GRASS-GIS environment. Various GRASS routines were used to obtain the terrain indices (Mitášová and Hofierka 1993). Elevation data were provided in a digital format by IGM (Military Geographic Institute of Italy) as 25-m grid-based DEM. The primary terrain indices employed in this study are: elevation (cm), distance from the middle stream (cm), slope gradient (degrees), slope aspect (degree counterclockwise from East), downward or upward flow-path lengths (cm), specific contributing area ($m^2$ $m^{-1}$), profile curvature, and plane curvature. We used the following secondary attributes: wetness index, stream power index, sediment transport index, and direct solar radiation for a given day of the year (Moore et al. 1993). Only correlations and relationships between topographic indices and soil variables measured at topsoil were explored.

### Site description and measurements

The study area was "Fiumarella di Corleto" river subcatchment (30 $km^2$) located about 50 km SW from the city of Potenza, (Italy). The parent material mainly consisted of clayey components with soils ranging from Vertisols to Mollisols, Inceptisols and Entisols, according to Soil Taxonomy. The environment had a very dynamic geomorphology and very interesting features from the soil-landscape modeling viewpoint (Santini et al. 1999). Two transects were established at the opposite sides of the Fiumarella river through distinct soil types (Santini and Romano 1999). Undisturbed soil cores were collected along the transects from the uppermost soil horizon at 88 sites, spaced 50 m apart. The soil cores were subjected to laboratory measurements to determine the particle-size distribution, organic carbon content, and soil water retention characteristics. The cores were oven-dried at 105 °C to determine dry bulk density. Total porosity was calculated from the measured oven-dry bulk density and particle density set at 2.65 $g/cm^3$. Organic carbon in soil was determined with the dichromate method. Sand, silt, and clay contents were expressed as percentage by mass of the fine-earth fraction (<2 mm) and soil texture was identified according to the USDA soil classification. Drying water retention data points θ(h) were measured at several pressure heads using a suction table apparatus (Romano et al. 2002). To allow comparisons between the measured and PTF-predicted retention curves, for each soil sample the measured θ(h) values were fitted with the van Genuchten's (van Genuchten et al. 1991) closed-form retention relation (hereinafter referred to as the VG model):

$$\theta(h) = \theta_r + (\theta_s - \theta_r)\left[1 + |\alpha h|^n\right]^{-m} \qquad (2)$$

where $\theta_s$ is the saturated water content and $\theta_r$ represents residual water content, whereas the VG parameters α, n, and m control the shape of the water retention curve. By imposing the condition m=1-1/n, Eq. (2) is defined by the unknown parameters $\theta_s$, $\theta_r$, α, n, which was estimated using the RETC software package (van Genuchten et al. 1991). For all of the collected soil cores the VG model fitted the available measured retention data very closely, and therefore the fitted water retention curves do represent a consistent and relatively unbiased information for comparison purposes.

## Method description

The soil water retention function $\theta(h)_{VG}$ can be interpreted as a probabilistic process:

$$\theta(h)_{VG} = \theta(h)_\tau + \varepsilon(h) \qquad (3)$$

where $\theta(h)_{VG}$ is described by Eq. (1) and fitted to the measured soil water retention data points. This function represents the observed soil water retention property at one generic position of the study transects. We assume for $\theta(h)_{VG}$ an additive model with a deterministic component $\theta(h)_\tau$, defined by a pedotransfer rule (for example, the PTF-VER proposed by Vereecken et al. 1989) together with terrain information, and a stochastic component ε(h) accounting of both random and systematic errors since both measurement errors and model errors contribute to produce noises in the data. The residual ε(h) can be spatially uncorrelated (white noise) or may exhibit explicitly some degrees of spatial correlation. The method employed in this paper to incorporate landscape features into pedotransfer predictions supposes that the residual between $\theta(h)_\tau$ and the original PTF-predicted soil water retention function, $\theta(h)_{PTF}$, is modeled by the following polynomial expression:

$$\theta(h)_\tau - \theta(h)_{\text{PTF}} = \sum_{j=1}^{t} a_j \tau_j \tag{4}$$

where $\tau_j$ denotes independent variable representing the generic terrain attribute (for example, slope or plain curvature) and $a_j$ is the coefficient of variable $\tau_j$. Specifically, as the soil water retention function (Eq. 2) is described by the set of four parameters $\theta_s$, $\theta_r$, $\alpha$, and n, we pose:

$$\begin{cases} \theta_{s,\tau} = \theta_{s,\text{PTF}} + \sum_{j=1}^{t} a1_j \tau_j \\ \theta_{r,\tau} = \theta_{r,\text{PTF}} + \sum_{j=1}^{t} a2_j \tau_j \\ \alpha_\tau = \alpha_{\text{PTF}} + \sum_{j=1}^{t} a3_j \tau_j \\ n_\tau = n_{\text{PTF}} + \sum_{j=1}^{t} a4_j \tau_j \end{cases} \tag{5}$$

where the subscript PTF means that the specific retention parameter is calculate using the pedotransfer algorithm as originally proposed by their authors. The unknown coefficient values appearing in Eq. 5 are determined by minimizing a performance-based objective function.

## Results and Discussion

The proposed procedure is evaluated by comparing selected, fitted water retention points with respect to the related PTF-predictions, as computed by including or not the landscape information.

The measure of prediction performance is computed in terms of integral mean deviation (IMD) and integral root-mean-square deviation (IRMSD), defined as follows:

$$\text{IMD} = \frac{1}{(\xi_u - \xi_l)} \int_{\xi_l}^{\xi_u} \left[\theta(\xi)_{\text{VG}} - \theta(\xi)_{\text{PTF}/\tau}\right] \mathrm{d}\xi \tag{6}$$

$$\text{IRMSD} = \left[\frac{1}{(\xi_u - \xi_l)} \int_{\xi_l}^{\xi_u} \left[\theta(\xi)_{\text{VG}} - \theta(\xi)_{\text{PTF}/\tau}\right]^2 \mathrm{d}\xi\right]^{\frac{1}{2}} \tag{7}$$

where $\xi = \log(|h|)$, $\theta(\xi)_{VG}$ is the observed soil water retention function, and $\theta(\xi)_{PTF/\tau}$ is the corresponding PTF prediction including or not terrain attributes. We integrated the soil water retention functions between the lower limit $\xi_l = 0.00$ and the upper limit $\xi_u = 4.20$. The IMD index reveals the presence of biases in prediction (systematic underprediction, IMD>0, or systematic overprediction, IMD<0), whereas IRMSD provides a measure of the overall precision, namely the degree of dispersion offered by a prediction method. If small systematic errors occur (i.e., small IMD), the prediction method is accurate; if IRMSD is small, the prediction method has good precision.

Table 1 presents the arithmetic means of percent IMD and IRMSD for the whole data set (N=88) for PTF-VER, whether or not including topographic variables into the original pedotransfer rule. The negative values of IMD% indicate that systematic overpredictions, albeit of different magnitude, are detected with respect to the observed water retention functions.

**Table 1.** Meant of IMD% and IRMSD% for PTF-VER calculated from the whole set of soil water retention functions.

| Type of Predicition | IMD% | IRMSD% |
|---|---|---|
| a) Original PTF | - 2.770 | 4.02 |
| b) PTF + 1 terrain attribute | | |
| PTF + slope | - 0.024 | 3.21 |
| PTF + aspect | - 0.624 | 3.36 |
| PTF + profc | - 0.128 | 3.34 |
| PTF + kw | - 0.069 | 3.31 |
| PTF + wi | - 0.411 | 3.41 |
| c) PTF + 2 terrain attributes | | |
| PTF + (slope & aspect) | 0.005 | 3.18 |
| PTF + (slope & profc) | 0.074 | 3.17 |
| PTF + (slope & kw) | 0.206 | 3.13 |
| PTF + (slope & wi) | 0.088 | 3.20 |
| PTF + (aspect & profc) | - 0.114 | 3.31 |
| PTF + (aspect & kw) | 0.006 | 3.18 |
| PTF + (aspect & wi) | - 0.281 | 3.30 |
| PTF + (profc & kw) | 0.064 | 3.20 |
| PTF + (kw & wi) | - 0.055 | 3.24 |

Terrain attributes: slope is slope gradient; aspect is slope aspect; profc is profile curvature; kw is solar radiation; wi is wetness index

Values of IMD close to zero means that on average there is little difference between observed and PTF-predicted soil water retention functions. While Tietje and Tapkenhinrichs (1993) and Romano and Santini (1997) reported an overall satisfactorily behavior of the pedotransfer published by Vereecken et al. (1989), for the data collected in the present study the original PTF-VER gives the worst result (data regarding other PTFs performance not presented here) in terms of mean IMD% value, with an average overprediction of 2.77%. In terms of IRMSD% the original PTF-VER shows overall a fair goodness-of-fit (mean IRMSD% is 4.02%), behaving slightly worse than PTF HYPRES of Wösten (1999) but fairly better than the PTF of Rawls and Brakensiek (1989). On the other hand, PTF-VER shows a fair sensitivity to inclusion of primary and secondary terrain attributes. Adding terrain attributes always yields improvement with respect to both bias and dispersion characteristics.

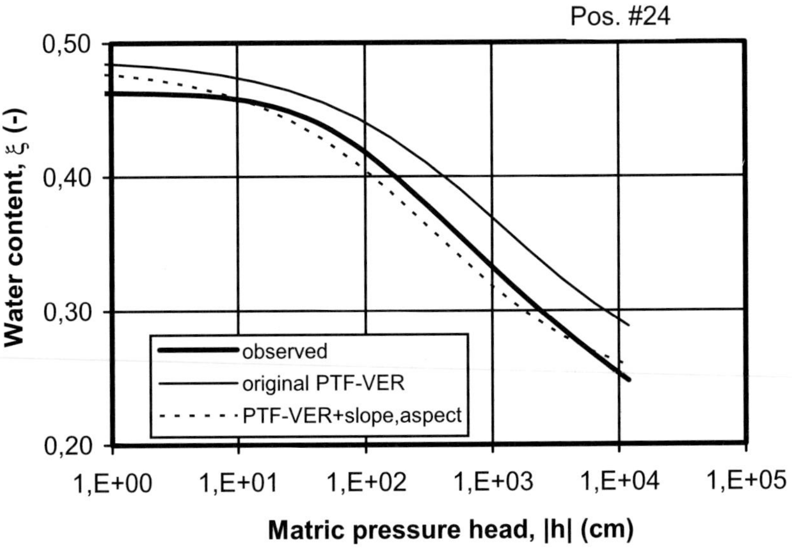


**Fig. 1.** Observed soil water retention for one location and comparison with the original, $\theta(h)_{PTF}$, or adjusted, $\theta(h)_{\tau}$ for PTF-VER

For PTF-VER, using both slope gradient and slope aspect substantially improves the average prediction of soil water retention. Slope gradient and slope aspect used together as ancillary information yield a virtually zero mean IMD% value (IMD% = 0.0055%) and this is an indication of highly unbiased mean water retention predictions. The benefit of considering various terrain attributes as ancillary variables does not appear significant in terms of the IRMSD% indices. For the case of PTF-VER+slope,kw,wi, we obtain the maximum change in the mean IRMSD% value of approximately -23.6% with respect to the original PTF-VER (i.e., from 4.02% to 3.07%). As illustrative examples, Fig. 1 examine the effects that refining a pedotransfer rule with the proposed method have on predictions of θ(h) functions. Specifically, with reference to the sampling point No. 24 located along the right-side transect, Fig. 1 shows the soil water retention characteristics as predicted by the original PTF-VER (thin line) or by adding the slope and aspect topographic attributes to this pedotransfer function (thin dashed line). The solid lines depict the observed θ(h) functions. The improvement in the description of soil water retention characteristic is evident.

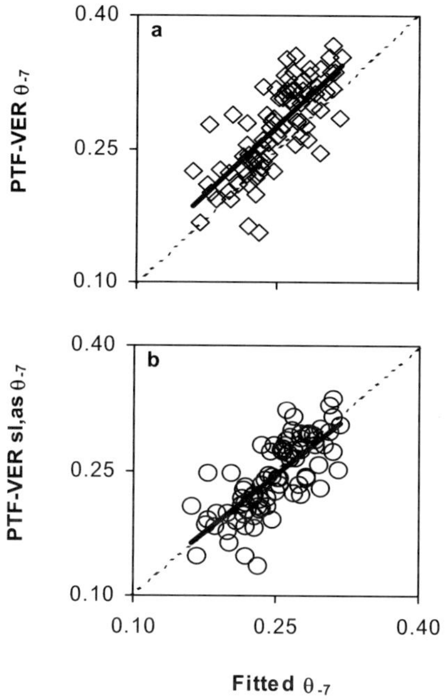


**Fig. 2.** Scatter plots of water content at 7 bar as measured and as predicted with PTF of Vereecken, original version (a) and including slope and aspect (b)

To reveal the magnitude of the enhanced prediction capability of PTFs when provided with topographic ancillary information, we produced also scatter plots of measured values of water content at selected values of matric potential versus PTF prediction, with and without topographic information embedded. For matric potential of 7 bar, Fig. 2 shows scatter plots of water content of all position along the transects as measured and as predicted with PTF of Vereecken, including (Fig. 2b) or not (Fig. 2a) slope and aspect as topographic ancillary information. The pedotransfer of Vereecken shows a light overprediction of water content (Fig. 2a), whereas with the inclusion of slope and aspect (Fig. 2b), linear interpolation curve is very close to the 1:1 line.

## Conclusions

This study has demonstrated the potential usefulness in improving PTF-predictions by accounting for landscape information. The research detailed in this paper has provided an approach for including terrain attributes into a pedotransfer rule to enable a better, but still relatively easy to achieve, description of the hydraulic behavior of soils in a region of interest. It is important to point out that we did not designed a new pedotransfer function. Rather, in this study attention was focused on how published, and already widely tested, PTFs can be suitably and inexpensively calibrated to account for site-specific situations. We have proposed to refine a pedotransfer function by a superposition of the original PTF and a linear combination of various terrain attributes that can be readily retrieved from a DEM. It is apparent that terrain attributes enhance the reliability of an original pedotransfer function and, more importantly, work efficiently to produce unbiased predictions. This refinement is expected to be more effective when addressing modeling problems at hillslope and small catchment scales. Both these features have valuable implications in distributed hydrologic modeling. Scaling up relies on some weighing techniques and thus the use of an unbiased prediction method is of crucial importance to compute reliable equivalent soil hydraulic parameters at the numerical grid scale. This can be not enough when running a comprehensive hydrologic model. A simple averaging of a certain variable could lead to a poor representation of the actual patterns of variability, especially if strong spatial correlations there exist, and one should make every endeavor to improve the description of soil spatial variability.

## References

Bastet G, Bruand A, Voltz M, Bornand M, Quétin P (1999) Performance of available pedotransfer functions for predicting the water retention properties of French soils. In Van Genuchten MT, Leij F, Wu L (eds.) Characterization and measurement of the hydraulic properties of unsaturated porous media; Proc. inter. workshop, Riverside (CA-USA), 22-24 October 1997, University of California, Riverside

Bouma J (1989) Using soil survey data for quantitative land evaluation. Adv Soil Sci 9:177-213

Burrough PA (1993) Soil variability: a late 20th century view. Soils Fert 56:529-562

Famiglietti JS, Rudnicki JW, Rodell M (1998) Variability in surface moisture content along a hill-slope transect: Rattlesnake Hill, Texas. J Hydrol 210:259-281

Mitášová H, Hofierka J (1993) Interpolation by Regularized Spline with Tension: II. Application to terrain modeling and surface geometry analysis. Mathematical Geology 25:657-669

Moore ID, Grayson RB, Ladson AR (1991) Digital terrain modelling: a review of hydrological, geo-morphological, and biological applications. Hydrol Processes 5:3-30

Moore ID, Gessler PE, Nielsen GA, Peterson GA (1993) Soil attribute prediction using terrain analysis. Soil Sci Soc Am J 57:443-452

Nemes A, Schaap, MG, Leij FJ (1999) The UNSODA unsaturated soil hydraulic database, version 2.0 U.S. Salinity Laboratory, USDA, ARS, Riverside, CA-USA

Rawls WJ, Brakensiek DL (1989) Estimation of soil water retention and hydraulic properties. In Morel-Seytoux HJ (ed.) Unsaturated flow in Hydrologic Modeling - Theory and Practices, NATO ASI Series. Dordrecht: Kluwer Academic

Refsgaard JC (1997) Parameterisation, calibration and validation of distributed hydrological models. J Hydrol 198:69-97

Romano N, Santini A (1997) Effectiveness of using pedotransfer functions to quantify the spatial variability of soil water retention characteristics. J Hydrol 202:137-157

Romano N, Hopmans JW, Dane JH (2002) Water retention and storage: Suction table. In: Dane JH and Topp GC (eds) Methods of Soil Analysis, part 4, Physical Methods, Soil Sci Soc of Am, Madison, Wis, pp. 692-698

Santini A, Romano N (1999) Medalus III (Mediterranean Desertification and Land Use), Project 1: Core Project. Final Report. Vol. 1 CEC, Brussels

Santini A, Coppola A, Romano N, Terribile F (1999) Interpretation of the spatial variability of soil hydraulic properties using a land system analysis. In Feyen J and Wiyo K (eds.) Modelling of transport processes in soils, Proc. inter. workshop, Leuven (Belgium), 24-26 Nov 1999, Wageningen: Wageningen Pers

Tietje O, Hennings V (1996) Accuracy of the saturated hydraulic conductivity prediction by pedotransfer functions compared to the variability within FAO textural classes. Geoderma 69:71-84

Tietje O, Tapkenhinrichs M (1993) Evaluation of pedotransfer functions. Soil Sci Soc Am J 57:1088-1095

van Genuchten MTh, Leij FJ, Yates SR (1991) The RETC code for quantifying the hydraulic functions of unsaturated soils. USEPA Rep. IAG-DW12933934, R.S. Kerr Environ. Res. Lab, U.S. Environmental Protection Agency, Ada, OK, USA

Vereecken H, Maes J, Darius P (1989) Estimating the soil moisture retention characteristic from texture, bulk density and carbon content. Soil Sci 148:389-403

Wösten JHM, Lilly A, Nemes A, Le Bas C. (1999) Development and use of a database of hydraulic properties of European soils. Geoderma 90:169-185

# Analytical Model for Gravity-Driven Drainage

G. Severino, A. Comegna, A. Sommella

Division of Water Resources Management, Naples University, ITALY

## Abstract

We model drainage flow by assuming gravitational regime. Central for the present paper is a fairly general analytical solution which is applicable to local (typically laboratory) scale. These results are then specified by adopting the Brooks et al (1964) model for the hydraulic conductivity curve.

At field scale, we combine our analytical results with the stream tube model of Dagan et al. (1979) to derive the ensemble average of the effective saturation $\langle S \rangle$. This is achieved by regarding the parameters of hydraulic properties as *random space functions* (RSFs) with given joint density distribution. For illustration purposes, we limit to consider the spatial variability of the saturated conductivity $K_s$, solely.

## Keywords

drainage; analytical solutions; heterogeneity; stochastic modelling

## Introduction

Concerns over groundwater recharge, irrigation efficiency, runoff, and transport of contaminants, have determined a great interest associated with a good understanding of flow processes occurring in the unsaturated zone. Usually, several parameters are needed to determine the hydraulic properties (i.e. retention and conductivity curves) which control such hydrological processes. A direct estimation of hydraulic properties is relatively difficult. More precisely, while the retention curve may be easily measured, determining the conductivity curve is not straightforward. For this reason, during the past decade many authors have proposed to estimate the conductivity by means of *ad hoc* drainage experiments (a recent review may be found in Romano and Santini 1999). These procedures are applicable at local (typically laboratory) scales, where the hydraulic parameters can be regarded as constant. At larger (e.g. field) scales, hydraulic properties may vary by several order of magnitudes because of the heterogeneous nature of porous formations (see, for instance, Russo and Bouton 1992).

Common to the estimation (both at local and field scale) of the hydraulic parameters is a drainage process that is used in the context of a nonlinear least squares procedure for calibration purposes. Most of previous studies were based on numerical solution of Richards equation which is used to simulate the direct problem. Only recently (see e.g. Severino et al. 2003), simplified solutions of Richards equation have been proposed to address the identification problem by means of analytical tools. The use of analytical solutions makes it possible to speed up the optimization procedure as well as to obtain more accurate calibrations.

In the present paper, we model drainage in the unsaturated zone. More precisely, we present a fairly general method to derive analytical solutions for a gravity-driven flow, and subsequently specify our results by adopting the Brooks and Corey (1964) model for the hydraulic conductivity. Under ergodic conditions, this solution is then averaged according to the methodology of the Dagan and Bresler (1979) column model to derive the field scale flow variables.

## General Results

We consider a drainage flow subsequent to the intrusion of a certain amount of water which determines a constant water content (say $\theta_W$) from the soil surface (corresponding to z=0) till to a given depth (say $z=z_W$). For simplicity, it is assumed that beneath $z_W$ the water content $\theta$ is equal to its irreducible value, i.e. $\theta=\theta_r$. Drainage occurs with zero flux at the surface, and we assume that flow is controlled by the gravitational forces, solely. Under these assumptions, flow is governed by the following equations

$$q \approx K(\theta) \qquad \frac{\partial \theta}{\partial t} + \frac{\partial q}{\partial z} = 0 \tag{1}$$

(the vertical coordinate $z$ is positive downwards). Several models expressing the dependence of $K$ upon $\theta$ are available in the literature (e.g. Gardner 1958, Brooks and Corey 1964, Mualem 1976, van Genuchten 1980), and they commonly assume that:

$$K(S)=K_sK_r(S) \tag{2}$$

where $K_r$ and $K_s$ represent the relative and saturation conductivity, respectively. The saturation S appearing into (2) is defined as $S=\dfrac{\theta-\theta_r}{\theta_s-\theta_r}$, being $\theta_s$ the saturated water content. Based on the shape of the hydraulic conductivity curve, $K_r$ is assumed to be an increasing function of S with $\ddot{K}_r>0$ (the "dot" notation represents the derivative with respect *S*). Equations (1) can be rearranged into a single first order nonlinear partial differential equation

$$\frac{\partial S}{\partial t}+\frac{K_sK_r}{\Delta}\frac{\partial S}{\partial z}=0 \qquad \Delta=\theta_s-\theta_r\,. \tag{3}$$

We solve (3) in a semi-infinite domain z>0 with the following initial and boundary conditions

$$S(z,0)=\begin{cases} S_w=S(\theta_w) & z<z_w \\ 0 & z>z_w \end{cases} \qquad \text{for } z>0,\quad S(0,t)=0 \text{ for } t>0. \tag{4}$$

The solution of problem (3)-(4) is achieved by the method of characteristics (see figure).

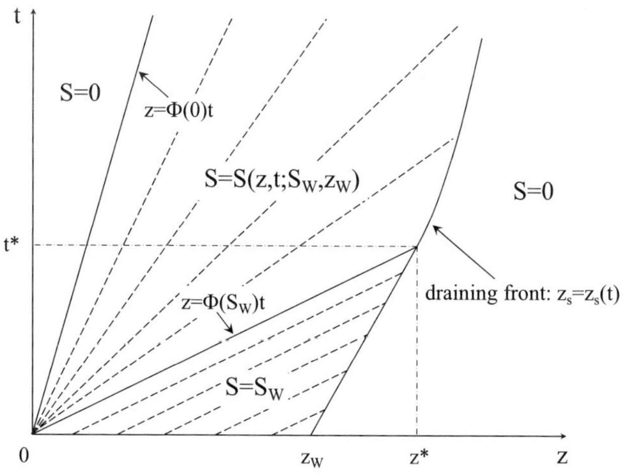


**Fig. 1.** Effective saturation in the characteristic plane

In the plane (z, t) the saturation S satisfies the ordinary differential equation $\frac{dS}{dt}=0$ provided that $\frac{dz}{dt}=\Phi(S)=\frac{K(S)}{\Delta}$. At z=0 we have 0= S(0, t)<$S_W$, and therefore the saturation "expands": starting from the origin there will be "fan-like" characteristics

$$z=\Phi(S)\,t \tag{5}$$

with slope Φ (S) ranging from Φ ($S_W$) to Φ (0). The resulting saturation is obtained from (5) as $S=\Phi^{-1}(z/t)$. Thus, for $0<z<z_W$, since S(z, 0)=$S_W$=const, the characteristics are parallel, and $S=S_W$ (see figure). For $z > z_W$, because $\Phi(0)<\Phi(S_W)$ the characteristic starting from $z=z_W$ intersects at $(z^*, t^*)$ the one starting from the origin. As time increases, the distance between $S=S_W$ and S=0 decreases, and therefore the saturation is "compressing".

As a consequence, a discontinuity curve $z_s=z_s(t)$ (hereafter termed as "draining front") separating the zone where the solution jumps from S>0 to 0 is present. The equation for the draining front is determined by the Rankine-Hugoniot condition, i.e.

$$\frac{dz_s}{dt}=\frac{[K]^*}{[S]^*\Delta} \tag{6}$$

where we have introduced the notation $[K]^*$ and $[S]^*$ for the jumps in K and S, respectively.

For given $t<t^*=\frac{z_W\Delta}{K(S_W)}\left[\frac{\dot{K}_r(S_W)}{K_r(S_W)}-\frac{1}{S_W}\right]^{-1}$, the saturation S jumps from $S_W$ to 0, and the draining front is the straight line $z_s=z_W+\frac{K(S_W)t}{S_W\Delta}$.

At $t>t^*$, since the saturation S is reducing, $z_s$ is not anymore a straight line and it is calculated from (5)-(6) as

$$\frac{dz_s}{dt}=\frac{K(S)}{S\Delta} \qquad \frac{z_s}{t}=\Phi(S). \tag{7}$$

Elimination of S in the system (7) yields the following initial value problem

$$\frac{dz_s}{dt} = \Psi\left(\frac{z_s}{t}\right) \quad z\left(t^*\right) = z^* = z_w\left\{1 + \frac{1}{S_w}\left[\frac{\dot{K}_r(S_w)}{K_r(S_w)} - \frac{1}{S_w}\right]^{-1}\right\} \tag{8}$$

being

$$\Psi\left(\frac{z_s}{t}\right) = \frac{K\left[\Phi^{-1}(z_s/t)\right]}{\Delta\Phi^{-1}(z_s/t)}. \tag{9}$$

Beyond (z*,t*) the S-profile overtakes the shock from behind so that the saturation on the left side of the shock decreases, while S=0 on the right. Furthermore, as $\frac{K(S)}{S}$ is reducing with the time, the draining front is continuously decelerating (see the first eq. of (7)). Finally, from the second of (8) we recover the following existence condition for $z_s(t)$:

$$\dot{K}_r > \frac{K_r}{S} \tag{10}$$

that is ensured since $\ddot{K}_r > 0$.

So far, we have dealt with a drainage process by regarding the soil as an homogeneous medium. These results are primarily applicable at laboratory scales. At larger (say field) scales one has also to account for the complicated structure of the flow paths (an exhaustive exposition can be found in the monographs of Gelhar 1993, and Zhang 2002). Usually, the horizontal extents of a field are characterized by a length scale which may vary quite largely. In many applications, we are mainly concerned with depths much smaller than the characteristic horizontal length scales, so that the medium can be considered vertically homogeneous. However, unlike the laboratory scale, we have now to account for the horizontal variations of hydraulic properties which are very irregular (e.g. Russo and Bouton 1992, Russo et al. 1997), and influence flow to a tremendous extent (see e.g. Bresler and Dagan 1979). We have shown that the saturation S (and similarly for the other flow variables) depends upon several quantities like: the hydraulic parameters ($K_s$, $\theta_s$, $\theta_r$, $\beta$), and initial saturation data ($z_W$, $S_W$ ). In order to calculate the spatial average $\overline{S}$ of the saturation, we use the concept of bundle of isolated columns proposed by Dagan and Bresler (1979). More precisely, $\overline{S}$ is sought as ensemble average by regarding the various parameters as random variables, and assuming ergodic conditions (Dagan and Bresler 1979), i.e.

$$\overline{S}(z,t) \approx \langle S(z,t)\rangle = \int d\alpha\, f(\alpha) H(t-t^*) H[z_s(t)-z] H[z-\Phi(0)t]$$
$$S(z,t) + \int d\alpha\, f(\alpha) H(t-t^*) \{ S_w H[z-\Phi(S_w)t] H[z_s(t)-z] \tag{11}$$
$$+ S(z,t) H[z-\Phi(0)t] H[\Phi(S_w)t - z] \}$$

being $f$ the joint probability distribution of the parameters ($K_s$, $\theta_s$, $\theta r$, $\beta$, $z_W$, $S_W$ ) $\equiv \alpha$. The mean saturation $\langle S\rangle$ is expressed by the aid of three contributions whose physical meaning is straightforward. The first term accounts for the columns where the saturation is below the initial saturated value $S_W$. The second represents the contribution of those columns which were initially saturated, and are not still reached by the draining front, whereas the last term of (11) accounts for the contribution of the upper columns in which the saturation has been already reduced. While at local scale the spreading of the saturation profile is due to the reduction of the conductivity, in the heterogeneous collection of columns there will also be a "stochastic spreading" due to the average effect upon all the columns.

The large times mean saturation $\langle S_\infty\rangle$ is obtained as

$$\langle S_\infty(z,t)\rangle \approx \int d\alpha\, f(\alpha) S(z,t). \tag{12}$$

Like the mean saturation, we can define the mean draining front $\langle z_s(t)\rangle$ as ensemble average over all the realizations of $z_s$, i.e.

$$\langle z_s(t)\rangle \approx \int d\alpha\, f(\alpha) z_S(t). \tag{13}$$

The knowledge of (13) is important especially when designing proper field scale experiments aiming to characterize the hydraulic properties of heterogeneous soils.

In general, most of the parameters (like $\theta_s$, $\theta_r$, $\beta$, $z_W$) appearing into (11) exhibit a small variability as compared with that of the others, and therefore they may be considered constant over the field (e.g. Indelman et al. 1998, Lessoff and Indelman 2002, Severino et al. 2003). Thus, based on such grounds and toward simplifying the computational aspect, we shall focus hereafter on the variability of only one parameter, namely the saturated hydraulic conductivity $K_s$, since drainage (and more generally flow) is highly sensitive to it (e.g. Russo 1993, Severino et al. 2003). As a consequence, the joint probability distribution function appearing in (11) is replaced by the probability distribution function of $K_s$.

## Application

We wish to specify the general results of the previous section in the case of the Brooks and Corey (1964) model, i.e.

$$K_r(S) = S^{\frac{1}{\beta}} \tag{14}$$

being β a parameter dependent upon the pore-size distribution. Furthermore, without lost of generality (and in line with the current standards of the experimental procedures) we shall assume that the soil is initially saturated up to $z_W$, i.e. $S_W = 1$. Under such assumptions, the equation for the draining front is easily calculated from (8)

$$z_s(t) = \begin{cases} z_w + \dfrac{K_s t}{\Delta} & t < t^* \\ z^* \left(\dfrac{t}{t^*}\right)^{\beta} & t > t^* \end{cases} \tag{15}$$

with $z^* = \dfrac{z_W}{1-\beta}$, and $t^* = \beta\Delta \dfrac{z^*}{K_s}$. The S-profile in the expansion zone writes as:

$$S(z,t) = \left(\frac{\beta \Delta z}{K_s t}\right)^{\frac{\beta}{1-\beta}}. \tag{16}$$

From (15) it is seen that at large times the length of the saturation profile grows like $t^\beta$, whereas the intensity (see (16)) decays with time as $t^{-\beta/(1-\beta)}$. Furthermore, we observe that, even if the initial saturation depth $z_W$ does not affect the value of the saturation (see (16)), it strongly influences the draining front $z_s=z_s(t)$, and the rate at which it advances. This has profound implications in both the optimal data-collecting (see Severino et al. 2003), and into the calibration procedure.

At field scale the mean value ⟨S⟩ is obtained from (11) as (we omit the algebraic derivations)

$$\langle S(z,t)\rangle = H(z_w - z)F(k_1) + H\left[z - z^*\Theta(k_2)\right]H(z^* - z)\Theta(k_1) + \\ + H(z - z_w)\left[F(k_1) - F\left(\beta\Delta\frac{z - z_w}{t}\right)\right] \tag{17}$$

being $F$(x) the cumulative distribution function, whereas

$$\Theta(a) = \int_a^\infty dK_s f(K_s) S(z,t), \quad k_1 = \beta\Delta \frac{z}{t}, \quad k_2 = \beta\Delta \left( \frac{z}{z_*} \right)^{\frac{1}{\beta}}. \tag{18}$$

In a similar manner the mean front is calculated as

$$\langle z_s(t) \rangle = z_w F(k) + \frac{t}{\Delta} \int_0^k dK_s K_s f(K_s) + \frac{z^*}{k^\beta} \int_k^\infty dK_s K_s^\beta f(K_s) \tag{19}$$

(with k=βΔz*/t). The mean saturation as well as $\langle z_s(t) \rangle$ can be computed explicitly once the shape of the probability distribution function $f$ is specified.

## Conclusions

In the present paper we have solved the drainage flow by means of analytical tools. This has been achieved by adopting the assumption of gravitational regime. In order to obtain the mean value of the effective saturation at field scale, we have combined the solution valid at local with the column model of Dagan and Bresler (1979). The mean saturation is expressed via a single quadrature by assuming that the saturated hydraulic conductivity is the only random variable.

Our analytical results may serve as tool for many applications, like prediction of the fate of chemicals in agricultural soils as well as groundwater recharge.

## References

Bresler E, Dagan G (1979) Solute Dispersion in Unsaturated Heterogeneous Soil at Field Scale: Applications. Soil Science Society of America Journal, 43, pp 461-467

Brooks PM, Corey AT (1964) Hydraulic Properties in Porous Media, Hydrol Paper 3. University of Colorado Fort Collins

Dagan G, Bresler E (1979) Solute Dispersion in Unsaturated Heterogeneous Soil at Field Scale: Theory. Soil Science Society of America Journal, 43, pp 461-467

Gardner WR (1958) Some steady state solutions of unsaturated moinsture flow equations with application to evaporation from a water table. Soil Science, 85, pp 228-232

Gelhar LW (1993) Stochastic Subsurface Hydrology. Prentice Hall

Indelman P, Touber-Yasur I, Yaron B, Dagan G (1998) Stochastic analysis of water flow and pesticides transport in a field experiment. Journal of Contaminant Hydrology, 32, pp 77-97

Lessoff SC, Indelman P (2002) Identifying Soil and Transport Properties Using a Model of Infiltration-Redistribution Flow and Transport in The Unsaturated Zone. In Rubin et al. (Eds), Preserving the Quality of Our Water Resources, Verlag Berlin Heidelberg, pp 16-33

Mualem Y (1976) A new model for predicting the hydraulic conductivity of unsaturated porous media. Water Resources Research, 12, pp 513-522

Romano N, Santini A (1999) Determining soil hydraulic functions from evaporation experiments by a parameter estimation approach: Experimental verifications and numerical studies. Water Resources Research, 35, pp 3343-3359

Russo D, Bouton M (1992), Statistical Analysis of Spatial Variability in Unsaturated Flow Parameters. Water Resources Research, 28, pp 1911-1925

Russo D (1993) Stochastic Modeling of Macrodispersion for Solute Transport in a Heterogeneous Unsaturated Porous Formation. Water Resources Research, 29, pp 383-397

Russo D, Russo I, Laufer A (1997) On the spatial variability of parameters of the unsaturated hydraulic conductivity. Water Resources Research, 5, pp 947-956

Severino G, Santini A, Sommella A (2003) Determining the soil hydraulic conductivity by means of a field scale internal drainage. Journal of Hydrology, 273, pp 234-248

van Genuchten MT (1980) A closed-form equation for predicting the hydraulic conductivity of unsaturated soils. Soil Sci Soc Am J, 44, pp 892-898

Zhang D (2002) Stochastic Methods for Flow in Porous Media. Academic Press

# 3 Integration

# Hydrogeophysical characterization of subsurface solute transport at the Krauthausen test site: experiments and numerical modelling

H. Vereecken, A. Kemna, A. Tillmann, J. Vanderborght, A. Verweerd

Institute Agrosphere, Dept. of Chemistry and Dynamics of the Geosphere, Forschungszentrum Jülich GmbH, Jülich- Germany

Corresponding author: Forschungszentrum Jülich, 52425 Jülich-Germany
Tel: +0049 2461 61 6392 – fax: +0049 2461 61 2518 – e-mail: h.vereecken@fz-juelich.de

## Abstract

We demonstrate the potential of geophysical methods, in particular Electrical Resistance Tomography (ERT) in conjuction with tracer experiments, to derive hydraulic variables (e.g. velocity) and solute transport parameters (e.g. dispersivity) in subsurface environments. For this purpose, real tracer experiments were monitored using crosshole time-lapse ERT at reference planes at the Krauthausen test site. Numerical inversion of an ERT dataset obtained from a numerical tracer experiment in a heterogeneous 3D flow domain reveals that the 2D distribution of the bulk electrical conductivity field at selected reference planes can be recovered. In 2000, a first small-scale tracer experiment at the Krauthausen field site combined with ERT was performed to monitor the change in bulk electrical conductivity at a reference plane.

The imaged bulk electrical conductivity changes were transformed to local concentrations using an empirically derived relation with the tracer (bromide) concentration. The spatial variation of the stream tube local dispersion and stream tube velocity in the reference plane was then derived. This information can also be used to derive inversely the statistical properties (variance and correlation length) of the hydraulic conductivity field. In 2002 and 2003, a larger scale 'positive' and 'negative' tracer experiment was conducted with the tracer electrical conductivity larger and smaller, respectively, than the groundwater electrical conductivity. The negative tracer experiment was conducted to avoid downward movement of the plume such that it passes through the most sensitive part of the ERT-image plane.

To improve the imaging of solute concentrations, a new Magnetic-Electrical-Resitivity-Imaging-Technique (MERIT) is being developed which uses the additional information contained in the magnetic field strength. In a first step, this technique is being made available at column and lysimeter scale. In a further step, MERIT will also be developed for field scale applications.

## Keywords

Hydrogeophysical methods; ERT; Modeling; tracer; MERIT

## Introduction: Upscaling transport in heterogeneous aquifers

To describe inert solute transport in a porous medium, the convection dispersion equation is commonly used to represent pore scale transport processes at the continuum scale:

$$\theta \frac{\partial C}{\partial t} = \nabla \theta \mathbf{D}_{\mathbf{d}} \nabla C - \mathbf{q} \nabla C \tag{1}$$

where $\theta$ ($L^3\,L^{-3}$)is the volumetric water content, $C$ ($M\,L^{-3}$) the concentration, $\mathbf{D_d}$ ($L^2T^{-1}$) the local-scale dispersion tensor, and $\mathbf{q}$ ($L\,T^{-1}$) the water flux vector. The local-scale dispersion includes all diffusive and dispersive processes happening on scales smaller than the averaging scale of the water flow.

In a heterogeneous medium, $\theta$, $\mathbf{q}$, and $\mathbf{D_d}$ vary with location $\boldsymbol{x}$, and the functions $\theta\,(\boldsymbol{x})$, $\mathbf{q}(\boldsymbol{x})$ and $\mathbf{D_d}(\boldsymbol{x})$ can only be specified in a statistical framework and Eq. (1) is a stochastic PDE. As a consequence, concentrations can only be predicted at a certain time and location in a statistical framework. In other words, only a probability distribution of concentrations at a certain time and location can be predicted. In the upscaled transport equation, the flow velocity $\mathbf{q}$ and $\theta(\boldsymbol{x})$ are assumed to be deterministic at the larger scale. The effect on the transport process of smaller scale velocity variations due to small scale hydraulic conductivity variations is lumped in an effective dispersion, $\mathbf{D}$. The spatial moments of the solute plume concentrations or the temporal moments of solute fluxes form the basis for the definition of the effective dispersion. Two basically different approaches can be used to determine moments.

In the first-approach, moments of expected concentrations or fluxes are determined. Expected in this context refers to the average in all realisations of the flow or hydraulic conductivity field. The spatial/temporal centralized second moments of expected concentrations/fluxes $\overline{X_{ij}}(t)$, $\overline{\sigma_t^2}(\mathbf{x})$ are:

$$\begin{aligned}\overline{X_{ij}}(t) &= \int x_i x_j \langle c_x(\mathbf{x},t)\rangle \mathbf{dx} - \int x_i \langle c_x(\mathbf{x},t)\rangle \mathbf{dx} \int x_i \langle c_x(\mathbf{x},t)\rangle \mathbf{dx} \\ c_x(\mathbf{x},t) &= C(\mathbf{x},t) \Big/ \int C(\mathbf{x},t)\mathbf{dx}\end{aligned} \quad (2)$$

$$\overline{\sigma_t^2}(\mathbf{x}) = \int t^2 \langle c_t(\mathbf{x},t)\rangle dt - \left(\int t \langle c_t(\mathbf{x},t)\rangle dt\right)^2 ;\ c_t(\mathbf{x},t) = J_s(\mathbf{x},t) \Big/ \left\langle \int J_s(\mathbf{x},t)dt \right\rangle \quad (3)$$

where <y> represents the expected value of $y$ in all realisations of the hydraulic conductivity and flow fields. In real cases, transport is observed in one realisation of the conductivity/ velocity field. But, in second-order stationary fields and solute plumes with a wide lateral extent, the spatial moments of the plume and the temporal moments of the averaged fluxes across a reference plane approximate the moments of ensemble averaged plumes and fluxes. In this case, the moments $\overline{X_{ij}}(t)$ and $\overline{\sigma_t^2}(\mathbf{x})$ quantify the spatial spreading of a solute plume at time $t$ (Figure 1) and the temporal spreading of the solute flux across a reference surface at a distance $x_1$ from the injection surface, respectively, due to local variations in flow velocity. In the second approach, the expected values of the spatial moments of a locally injected solute plume, $\langle X_{ij}(t)\rangle$ or temporal moments of locally observed solute breakthrough curves $\langle \sigma_t^2(\mathbf{x})\rangle$ are determined:

$$\langle X_{ij}(t)\rangle = \left\langle \int x_i x_j c_x(\mathbf{x},t)\mathbf{dx} - \int x_i c_x(\mathbf{x},t)\mathbf{dx} \int x_i c_x(\mathbf{x},t)\mathbf{dx} \right\rangle \quad (4)$$

$$\langle \sigma_t^2(\mathbf{x})\rangle = \left\langle \int t^2 c_t(\mathbf{x},t)dt - \left(\int t c_t(\mathbf{x},t)dt\right)^2 \right\rangle \quad (5)$$

$\langle X_{ij}(t)\rangle$ characterizes the average spreading of a plume resulting from a point source injection around its center of mass, whereas $\langle \sigma_t^2(\mathbf{x})\rangle$ represents the average spreading of a locally observed breakthrough curve around the local peak arrival time (Figure 2). These moments quantify the mixing or dilution of mass that is injected in a heterogeneous aquifer. Apparent dispersivity lengths, $\lambda_{Leq}$ and $\lambda_s$, can be defined so that the moments predicted by an 'upscaled' or 'equivalent' convection dispersion equation in a homogeneous flow field correspond with those in the heterogeneous one:

$$\lambda_{Leq}(t) = \frac{\overline{X_{11}(t)}}{2vt} \text{ or } \lambda_{Leq}(x_1) = \frac{\overline{\sigma_t^2(x_1)}v^2}{2x_1} \tag{6}$$

$$\lambda_s(t) = \frac{\langle X_{ij}(t)\rangle}{2vt} \text{ or } \lambda_s(x_1) = \frac{v^2\langle\sigma_t^2(\mathbf{x})\rangle}{2x_1} \tag{7}$$

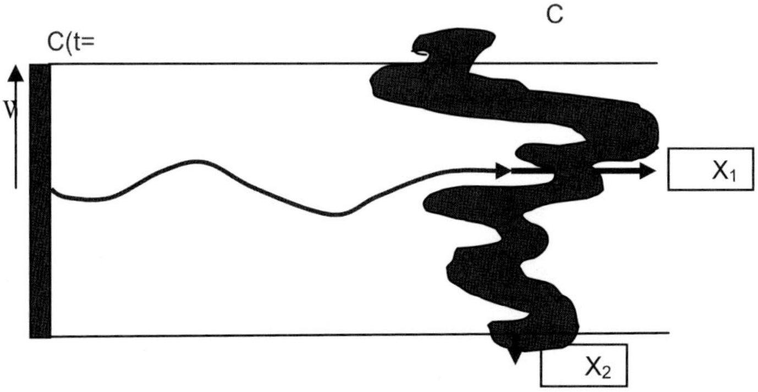


**Fig. 1.** Spatial central second moments, $X_{ij}(t)$, of a wide solute plume parameterize plume spreading due to spatial variability of flow velocities and correspond with those, $\overline{X_{ij}}(t)$, of expected concentrations in several realisations of the conductivity field

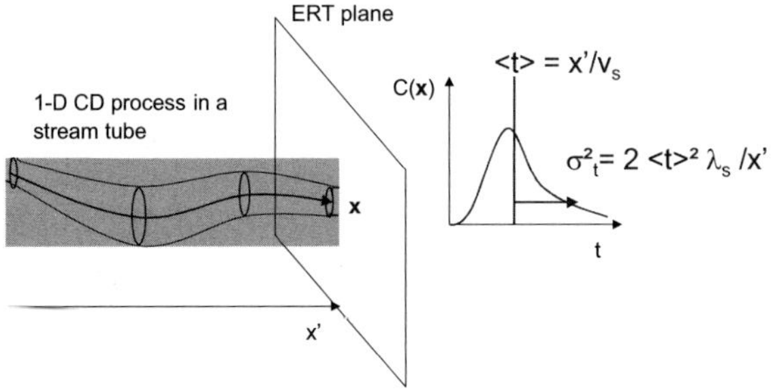


**Fig. 2.** Locally observed breakthrough curve (BTC) is interpreted as the result of a 1-D convection dispersion process in a stream tube. The time moments of the BTC are used to define stream tube parameters, $\lambda_s$ and $v_S$. The stream tube dispersivity $\lambda_s$ is a measure for the dilution and the spatial variability of the stream tube velocity, $v_S$, of the advection velocity variability

The equivalent dispersivity, $\lambda_{Leq} = D/v$, ($v$ is the mean velocity $<\mathbf{q}>/\theta$), describes the spatial spreading in the flow direction or the temporal spreading of the averaged solute fluxes. The dispersivity $\lambda_s$ represents the dispersivity of a 1-D convection dispersion process in a stream tube with a constant stream tube velocity, $v_s$, and describes the dilution of the local concentrations in the heterogeneous flow field. The stream tube velocity, $v_s$, is defined by the center of mass of the local plume or the average arrival time of a locally observed breakthrough curve (Figure 2).

The spatial heterogeneity of the $\log_e$ transformed hydraulic conductivity is represented by means of a spatial covariance function, $C_{\ln K}(\mathbf{h})$. For instance, an exponential covariance function can be used:

$$C_{\ln K}(\mathbf{h}) = \langle \ln K(\mathbf{x}) \ln K(\mathbf{x}+\mathbf{h}) \rangle - \langle \ln K \rangle^2 = \sigma^2_{\ln K} \exp\left[ -\sqrt{\frac{h_1^2}{\gamma_1^2} + \frac{h_2^2}{\gamma_2^2} + \frac{h_3^2}{\gamma_3^2}} \right] \tag{8}$$

where $\sigma^2_{\ln K}$ is the variance of ln$K$ and $\gamma_i$ the spatial correlation scale in direction i. By solving the stochastic transport equation Eq. (1), a direct relation between the structure of the aquifer heterogeneity, parameterized by $\sigma^2_{\ln K}$ and $\gamma_i$, the local scale dispersion, $\mathbf{D_d}$, and the equivalent parameters $\lambda_{\mathrm{eqL}}$ and $\lambda_s$ is obtained. Approximate analytical solutions of the stochastic flow and transport equations are obtained by expanding the solution of the partial differential equation in an asymptotic series of the input parameter perturbations. For small perturbations the asymptotic series may be truncated to obtain closed form approximate analytical solutions (e.g. Vanderborght and Vereecken, 2002). When the structure of the aquifer heterogeneity and local scale processes are parameterized, the equivalent parameters may be derived.

As an alternative, these parameters may be derived from tracer experiments. Finally, the structure of the aquifer heterogeneity may be inferred inversely from tracer experiment data. In the paper we first discuss the interpretation of point-scale concentration observations during tracer experiments that were carried out at the test-site Krauthausen (Germany). In the second part, the use of non-invasive imaging techniques to obtain spatially continuous information of subsurface transport processes is illustrated. How this information can be used for inverse characterization of aquifer heterogeneity is discussed.

## Point-scale monitoring of tracer experiment

At the test-site Krauthausen (Vereecken et al., 2000) multi-level groundwater samplers (MLS) were installed. At each well, the groundwater can be locally sampled at 24 depths between 3 and 10 m below the soil surface. The movement of a $Br^-$ tracer during a tracer experiment was monitored with the MLS. In Figure 3, the spatial extent of the solute plume, derived from spatial interpolation of concentration measurement in the MLS, is shown together with the breakthrough curves observed in different local samplers that are located at approximately the same distance from the injection wells. Both plots illustrate a large spatial variability of the transport process. Especially to capture the spatial extent and inner structure, a dense network of MLS is required. From the spatial distribution, the spatial moments were calculated and effective dispersion coefficients, $\lambda_{Leq}$, were derived (Vereecken et al., 2000). The breakthrough curves (BTCs) in different MLS at similar distances from the injection wells were grouped and averaged from which $\lambda_{Leq}$ were derived (Vanderborght and Vereecken, 2001). Stream tube dispersivities, $\lambda_s$, were derived from locally measured BTCs (Figure 4).

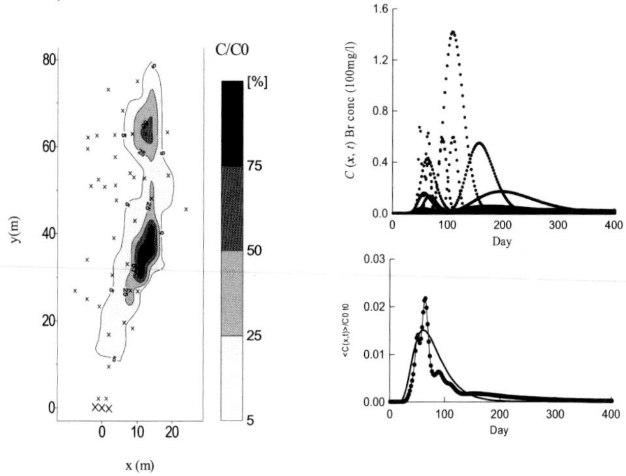


**Fig. 3.** Left panel: spatial extent of the $Br^-$ tracer plume at the test site Krauthausen at 85 days after plume injection (small crosses: multi-level samplers (MLS), large crosses: tracer injection wells). Right panels: local BTCs (top) and averaged BTC (bottom) in MLS at 55 m from the injection wells

The heterogeneity of the hydraulic conductivity at the test-site Krauthausen was derived from sediment analyses of sediment cores from the locations where the MLS were installed (Vereecken et al., 2000). The parameters $\sigma^2_{\ln K}$ and $\gamma_i$ derived from the experimental covariogram were uncertain because the hydraulic conductivity was derived indirectly from grain-size distributions and the horizontal distance between two MLS was relatively large. A comparison between first-order estimates of $\lambda_{Leq}$ based on the experimentally derived $\sigma^2_{\ln K}$ and $\gamma_i$ and $\lambda_{Leq}$ derived from averaged BTCs showed a relatively good agreement (Figure 4). The first-order approximate relations show that $\lambda_{Leq}$ or spreading is relative insensitive to the local scale dispersion, $\mathbf{D_d}$. The stream tube dispersivity, $\lambda_s$, is strongly determined by the local dispersion $\mathbf{D_d}$, especially by the transverse component of the local dispersion tensor.

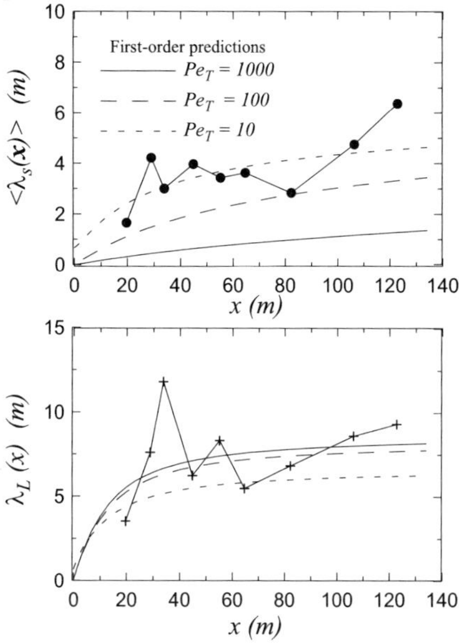


**Fig. 4.** stream tube, $\lambda_s$, (top), and equivalent dispersivity, $\lambda_{Leq}$ (bottom). (symbols: estimates from measured BTCs; lines: first-order approximate predictions for different transverse Peclet numbers: $Pe_T = \gamma_3/\lambda_{dT}$ where $\gamma_3$ is correlation length in vertical direction and $\lambda_{dT}$ local scale transverse dispersivity)

From fitting the first-order expression to $\lambda_s$ derived from local breakthrough curves, Vanderborght and Vereecken (2002) derived the transverse local scale dispersion coefficient. In order to verify the first-order approximate estimates of $\lambda_{Leq}$ and $\lambda_s$, a numerical experiment in a generated heterogeneous aquifer was carried out using the 3-D flow and transport simulation code TRACE/PARTRACE (Vereecken et al.,1994). More details about the numerical experiment are given in section (3.1). A tracer was injected in a planar source perpendicular to the mean flow direction. Time moments of BTCs of averaged solute fluxes Eq. (3) and of local concentrations Eq. (5) in reference planes were calculated from which $\lambda_{Leq}$ Eq. (6) and $\lambda_s$ Eq. (7) were derived. Figure 5 illustrates a close agreement between $\lambda_{Leq}$ and $\lambda_s$ derived from the numerical experiment and from first-order approximations. The stream tube dispersivity, $\lambda_s$, which parameterizes mixing and dilution in a heterogeneous flow field, is considerably larger than the pore scale dispersion. As a consequence, due to the heterogeneous flow field, which distorts the solute plume, mixing is increased compared with mixing in a homogeneous flow field.

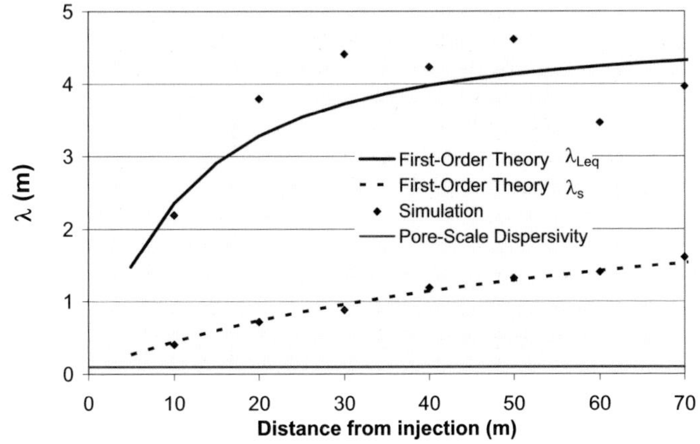


**Fig. 5.** stream tube, $\lambda_s$ and equivalent dispersivity, $\lambda_{Leq}$, derived from numerically simulated BTCs in a synthetically generated heterogeneous conductivity field (symbols) and from first-order approximate solutions (lines)

## Hydrogeophysical Imaging Using Electrical Resistivity Tomography (ERT)

Imaging of subsurface flow and transport processes using geophysical techniques represents an attractive approach for an improved characterization of many hydrological and environmental problems. In particular electrical resistivity tomography (ERT) has proven significant value in this regard if applied in a time-lapse manner, for instance to map water movement in the vadose zone (Daily et al. 1992) or to monitor solute transport in soils (Binley et al. 1996) and aquifers (Kemna et al. 2002).

In the following we demonstrate the potential of ERT for solute transport imaging by showing results from a synthetic experiment as well as from real field applications at the Krauthausen test site. For methodological aspects of time-lapse ERT as applied in these examples, including electrical modelling and inversion approaches, it is referred to Kemna et al. (2002) and Kemna et al. (2004).

## Synthetic Tracer Experiment

A synthetic model was set up to study the potential of 2D time-lapse ERT for quantitative imaging of 3D solute transport in a heterogeneous aquifer. The experimental design followed a typical field-scale application involving crosshole ERT. A 3D hydraulic conductivity field was generated assuming that the log-transformed hydraulic conductivity structure is represented as a second-order stationary random field (spatial stochastic process) with mean and variance values ($\sigma^2_{\ln K}=1$) and an exponential spatial covariance structure ($\gamma_1=\gamma_2 = 5$ m, $\gamma_3 = 1$ m). The employed realization in the considered domain is shown in Figure 6.

Imposing a hydraulic gradient in $y$ direction on the domain, and assuming constant porosity and constant longitudinal and transverse dispersivities ($\lambda_{dL} = 0.1$ m; $\lambda_{dT} = 0.01$ m), transport of a solute plume was modelled on the basis of the 3D convection-dispersion equation. Initially (day 0), a constant solute concentration was assumed in a rectangular region ($25 \leq x \leq 75\,\text{m}$, $5 \leq z \leq 15\,\text{m}$) at $y = 20$ m, and a zero concentration elsewhere.

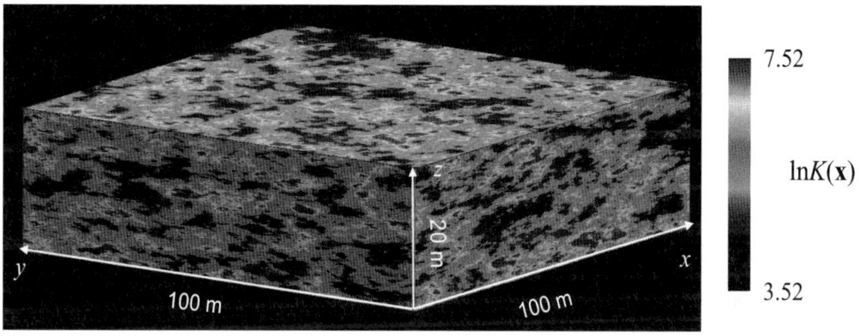


**Fig. 6.** 3D hydraulic conductivity field used for solute transport simulation

Down-gradient of the tracer "injection" plane, a fictive ERT image plane was defined at $y = 50$ m, comprising several equally separated boreholes, each equipped with electrodes. For the simulation of the ERT survey, a change in solute concentration was assumed to be linearly related to a change in bulk electrical conductivity (EC). Details on the simulation (e.g., data collection scheme, 3D electrical modelling, data noise) are given in Kemna et al. (2004).

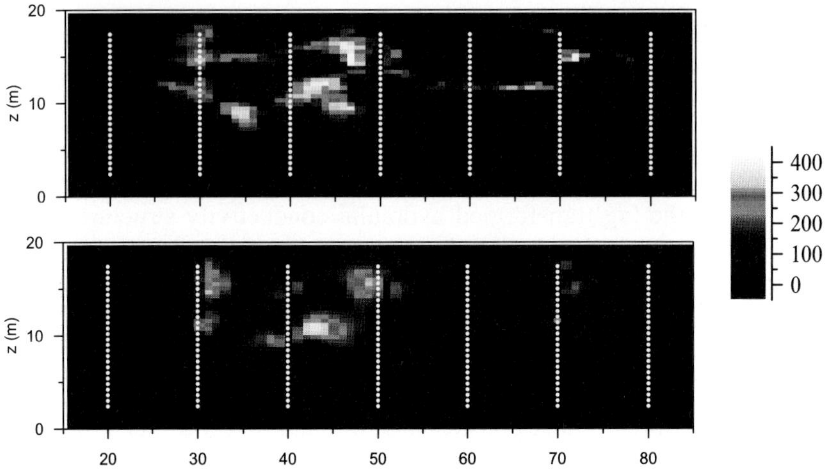


**Fig. 7.** Relative changes of electrical conductivity (in %) associated with the breakthrough of a synthetic solute plume in a cross-section perpendicular to the mean flow direction: Original distribution (top) and 2D ERT imaging result (bottom). Solid circles indicate the position of electrodes used for fictive 3D ERT data collection

Figure 7 shows an exemplary 2D ERT imaging result (day 31) together with the corresponding original distribution, demonstrating the general capability of the approach to reconstruct the solute plume in the image plane. Given the inherent resolution limits of ERT, the heterogeneity of transport is fairly well captured.

## Field-Scale Tracer Experiments

In order to investigate solute transport in heterogeneous aquifers, groundwater tracer experiments conducted at the Krauthausen test site in the years 2000, 2002 and 2003 were monitored using time-lapse ERT. Figure 8 provides an overview of the test site showing the position of monitoring wells and the two regions selected for the tracer tests.

In Sep-Dec 2000, a first study was carried out using 2,000 l of a NaBr solution with a concentration of 15 g/l as an electrically conductive tracer. The tracer was injected between 6 and 7 m depth in the filtered section of a monitoring well (at a rate of 500 l/h), and monitored down-gradient in an ERT image plane perpendicular to the anticipated mean flow direction (see Figure 8). The image plane was spanned by electrodes in three boreholes and additional electrodes at the surface.

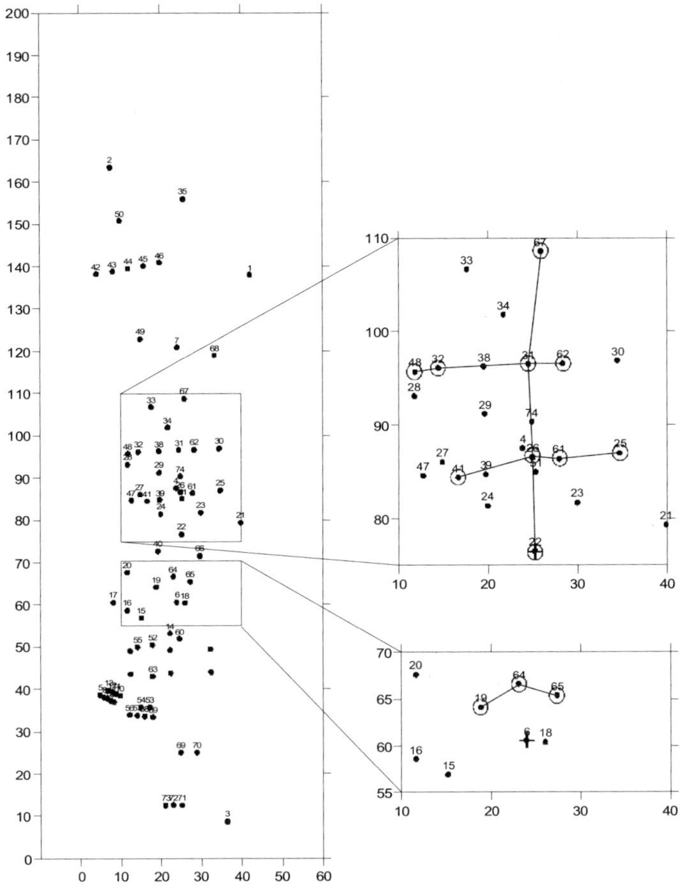


**Fig. 8.** Map view of the Krauthausen test site showing the position of monitoring wells. The two areas zoomed in on the right-hand side were selected for tracer experiments carried out in conjunction with ERT monitoring in the years 2000 (bottom) and 2002/2003 (top). Crosses indicate the position of respective injection wells; open circles indicate the position of ERT boreholes defining, respectively, one (2003) and three (2002/2003) vertical image planes

Although – mainly because of gravitational sinking of the tracer plume – solute breakthrough occurred in a less sensitive region of the ERT image plane, the principal part of the plume could be detected (Figure 9). Calibration relations between water EC and tracer concentration obtained from multi-level sampling results provided the basis for a quantitative analysis of transport characteristics (Kemna et al. 2002).

As a continuation of the smaller-scale tracer test conducted in 2000, two larger-scale experiments were carried out in the years 2002 and 2003, respectively. “Up-scaling” was accomplished by (i) injecting larger tracer volumes over longer time periods (140,000 l over 7 d), (ii) injecting tracer solution over almost the entire aquifer depth (between -3 to -10 m), and (iii) monitoring tracer breakthrough at crosshole reference planes further away from the injection well (see Figure 8).

In both experiments the same amount of tracer solution was injected in the same well and at the same time of the year (Sep-Dec) to ensure similar hydrological boundary conditions. In the first experiment (2002), a $CaCl_2$ tracer solution (2.9 g/l) with a higher EC than the groundwater was injected, whereas a “negative” tracer solution with a lower EC than the groundwater was injected in the second experiment (2003). During both tracer experiments, multi-level groundwater samples were taken to compare local groundwater EC measurements with ERT results. Calibration relations between bulk EC and groundwater EC were determined in the laboratory on columns packed with aquifer material.

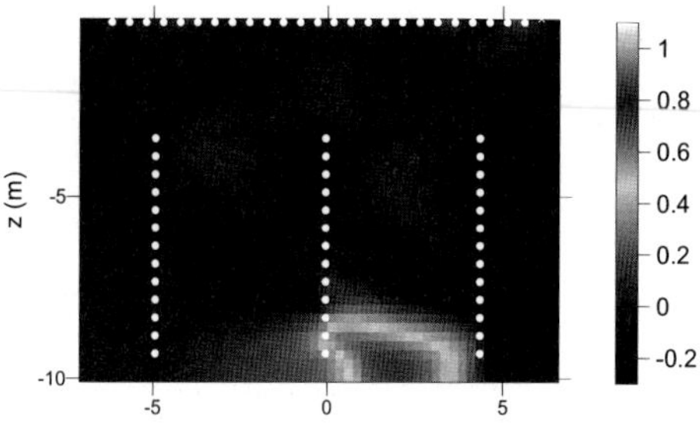


**Fig. 9.** ERT imaging result showing relative changes of electrical conductivity associated with the breakthrough of a NaBr tracer in a cross-section perpendicular to the mean flow direction at the Krauthausen test site. Solid circles indicate the position of electrodes used for ERT data collection

First results indicate a distinctly different behaviour of the two tracer plumes. A density driven downward drift was observed for the $CaCl_2$ plume which led to a more lateral spreading on top of the aquifer basis. First comparisons between ERT imaging results and multi-level sampler measurements show a good agreement (Figure 10).

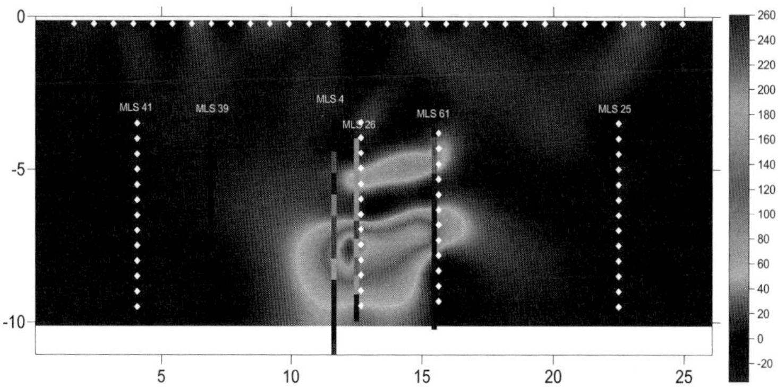


**Fig. 10.** ERT imaging result showing relative changes of electrical resistivity (in %) associated with the breakthrough of a "negative" tracer in a cross-section perpendicular to the mean flow direction at the Krauthausen test site. Solid rhombi indicate the position of electrodes used for ERT data collection; overlain colour bars show the corresponding results obtained from conventional multi-level sampling after application of proven calibration relations

## Interpretation of subsurface transport imaging of using stochastic models

The images obtained with ERT provide us with information on temporal and spatial variation of solute concentrations. The next important step is to parametrize subsurface transport from these data. In a first phase, we used data from the synthetic tracer experiment that was discussed in section 3.1 to demonstrate the feasibility of ERT for imaging and parameterizing subsurface transport.

From the simulated and ERT recovered tracer distributions, time moments of averaged and local BTCs were derived. Besides the equivalent dispersivities, $\lambda_{Leq}$ and $\lambda_s$, the spatial distribution of the stream tube velocities, $v_s$, was derived (Figure11). The structure of the spatial $v_s$ distribution was characterized by a spatial covariance function, $C_{vs}(\mathbf{h})$:

$$C_{v_s}(\mathbf{h}) = \langle v_s(\mathbf{x}) v_s(\mathbf{x}+\mathbf{h})\rangle - \langle v_s \rangle^2 \tag{9}$$

and was relatively well recovered by ERT. Using first- order approximate solutions of the stochastic flow and transport equation, $C_{vs}(\mathbf{h})$ can be predicted from $C_{\ln K}(\mathbf{h})$ and $\mathbf{D_d}$.

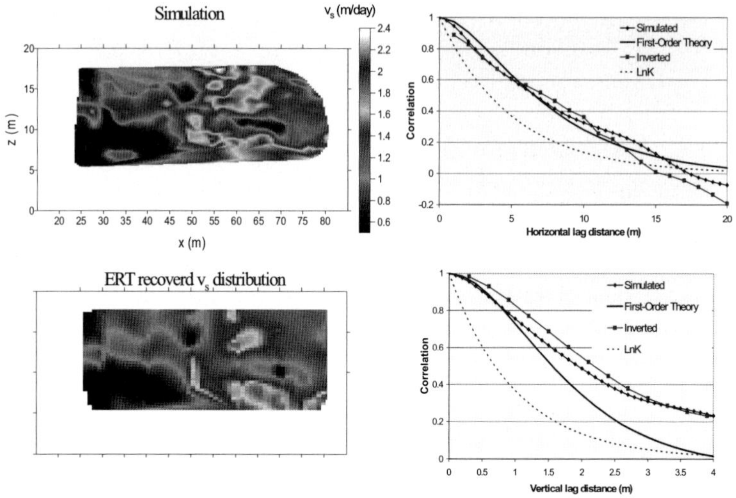


**Fig. 11.** Left panels: spatial distribution of stream tube velocities, $v_s$, on a reference plane perpendicular to the image plane derived from simulated (top) and ERT recovered BTCs (bottom). Right panels: spatial covariance of $v_s$ in horizontal (top) and vertical direction (bottom) derived from simulated (blue line) and ERT recovered BTCs (red line) and from first-order approximate solutions (full black line)

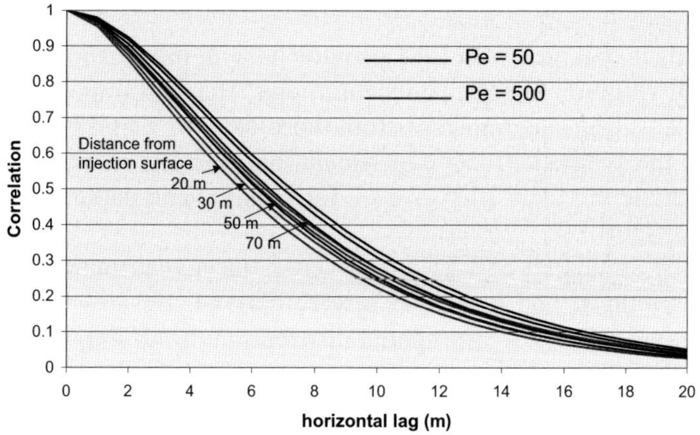


**Fig. 12.** First-order predictions of the spatial correlation of the stream tube velocity, $v_s$, in reference planes at different distances from the injection surface and for different local scale dispersions $D_d$ (Pe = $\lambda_{dL}/\gamma_I$)

Using first- order approximate solutions of the stochastic flow and transport equation, $C_{vs}(\mathbf{h})$ can be predicted from $C_{lnK}(\mathbf{h})$ and $\boldsymbol{D}_d$. A sensitivity analysis indicates that $C_{vs}(\mathbf{h})$ is relatively insensitive to the distance of the image plane from the injection surface or from the travel time and from the local scale dispersion (Figure 12). This suggests that the spatial correlation of the stream tube velocities is relatively robust against smearing due to local scale dispersion, so that the spatial correlation lengths, $\gamma_2$ and $\gamma_3$ of the hydraulic conductivity may be inferred inversely from the spatial correlation of $v_s$ without knowing $\boldsymbol{D_d}$. Since also $\lambda_{Leq}$ is relatively independent of $\boldsymbol{D_d}$ (Figure 4), the variance of $lnK$ may be inferred from $\lambda_{Leq}$ when it is assumed that $\gamma_1 = \gamma_2$. Finally, the local scale dispersion may be derived from $\lambda_s$. To get a better estimate of the parameters, this parameter estimation loop may be repeated using parameter estimates from the previous loop.

## Magneto-Electrical Resistivity Imaging Technique (MERIT)

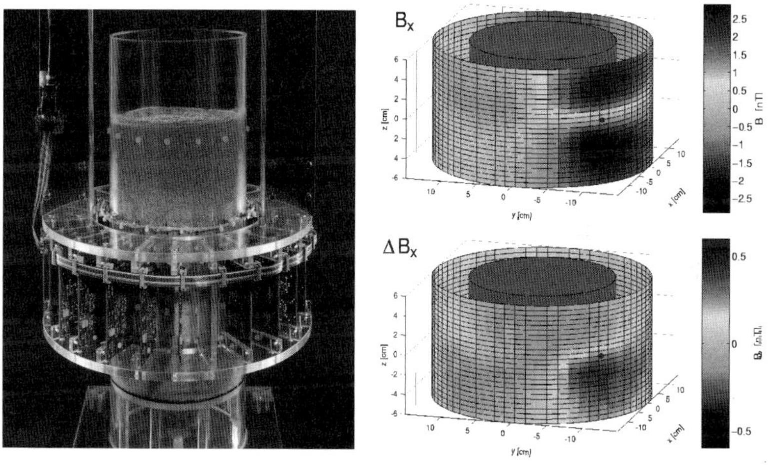


**Fig. 13.** Left: MERIT scanning system for imaging flow and transport in soil columns. Right: Magnetic field data on a dry sand column (top) and magnetic field changes due to partial water saturation (bottom) for a current injection between two diametrically positioned electrodes (red circles)

The novel Magneto-Electrical Resistivity Imaging Technique (MERIT) aims at combining conventional ERT with magnetic field measurements in order to improve the capabilities for flow and transport imaging. Recently, first imaging approaches for magnetic field data were presented (Kulessa et al. 2002, Kemna et al. 2003). Numerical studies proved the enhanced resolution of MERIT compared to standard ERT (Tillmann et al. 2003), and yielded information about magnetic sensor sensitivity and measurement configuration requirements. For the investigation of soil columns and lysimeters, a specifically designed MERIT system was developed (Zimmermann et al. 2004). The fully computer-controlled system consists of a vertically moving ring with 24 three-component anisotropic magnetoresistive sensors for magnetic sensing (Figure 13, left), and additional electrodes mounted on the object under investigation for electric current injection and potential measurement. Synthetic experiments proved the capability of the system to detect spatio-temporal water content variations (Figure 13, right) as well as structural heterogeneities (Tillmann et al. 2004).

## Conclusions

We presented the potential of combining Electrical Resistivity Tomography (ERT), tracer experiments and stochastic transport theories to quantify the hydraulic and transport properties of surface-near groundwater. Numerical ERT inversions using a simulated tracer experiment in a heterogeneous 3D flow field showed that the 2D distribution of the bulk electrical conductivity can be recovered at selected references planes. The approach was tested in a small scale tracer experiment at Krauthausen, 2000. We showed that the spatial variation of the stream tube local dispersion and velocity can be derived at selected reference planes using ERT and existing transport theories. The information obtained at the reference planes can be used to derive inversely the statistical properties of the hydraulic conductivity field. Two tracer tests were conducted in 2002 and 2003 to further test the potential of the presented approach. In addition the ERT method is further developed to MERIT (Magnetic-Electrical-Resistivity-Imaging-Technique) by simultaneously measuring potential differences and the 3D magnetic field strengths. We expect improved imaging of the solute concentrations in the aquifer.

## References

Binley A, Henry-Poulter S, Shaw B (1996) Examination of solute transport in an undisturbed soil column using electrical resistance tomography. Water Resources Res, 32:763-769

Daily WD, Ramirez AL, LaBrecque DJ, Nitao J (1992) Electrical resistivity tomography of vadose water movement. Water Resources Res 28:1429-1442

Kemna A, Vanderborght J, Kulessa B, Vereecken H (2002) Imaging and characterisation of subsurface solute transport using electrical resistivity tomography (ERT) and equivalent transport models. J Hydrology 267:125-146

Kemna A, Tillmann A, Verweerd A, Zimmermann E, Vereecken H (2003) MERIT – a new magneto-electrical resistivity imaging technique: 1) Modelling and tomographic reconstruction. Proc. 3rd World Congress on Industrial Tomography, Univ. of Calgary / VCIPT / UMIST / Univ. of Leeds, pp 256-261

Kemna A, Vanderborght J, Hardelauf, H, Vereeecken H (2004) Quantitative imaging of 3D solute transport using 2D time-lapse ERT: A synthetic feasibility study. Proc. Symp. Application of Geophysics to Engineering and Environmental Problems (SAGEEP), Environ Eng Geophys Soc pp 342-353

Kulessa B, Jaekel U, Kemna A, Vereecken H (2002) Magnetometric resistivity (MMR) imaging of subsurface solute flow – inversion framework and laboratory tests. J Environ Eng Geophys 7:111-118

Tillmann A, Verweerd A, Kemna A, Zimmermann E, Vereecken H (2003) Non-invasive 3D conductivity measurements with MERIT. Proc. Symp. Application of Geophysics to Engineering and Environmental Problems (SAGEEP), Environ Eng Geophys Soc pp 516-522

Tillmann A, Kasteel R, Verweerd A, Zimmermann E, Kemna A, Vereecken H (2004) Non-invasive 3D conductivity measurements during flow experiments in columns with MERIT: Proc. Symp. Application of Geophysics to Engineering and Environmental Problems (SAGEEP), Environ Eng Geophys Soc pp 618-624

Vanderborght J, Vereecken H (2001) Analyses of locally measured bromide breakthrough curves from a natural gradient tracer experiment at Krauthausen. J Contam Hydrology 48:23-43

Vanderborght J, Vereecken H (2002) Estimation of local scale dispersion from local breakthrough curves from a natural gradient tracer experiment at Krauthausen: the Langrangian approach. J Contam Hydrology 54:141-171

Vereecken H, Lindenmayr G, Neuendorf O, Döring U, Seidemann R (1994) TRACE – A mathematical model for reactive transport in 3D variably saturated porous media. KFA/ICG-4 Internal report No. 501494, pp 67

Vereecken H, Döring U, Hardelauf H, Jaekel U, Hashagen U, Neuendorf O, Schwarze H, Seidemann R (2000) Analysis of solute transport in a heterogeneous aquifer: The Krauthausen field experiment. J Contaminant Hydrology 45:329-358

Zimmermann E, Verweerd A, Glaas W, Tillmann A, Kemna A (2004) A new AMR sensor based magneto-electrical resistivity imaging system. IEEE Sensors, *submitted*

# Tracer Experiments on Field Scale for Parameter Estimation to calibrate Numerical Transport Models

J. Fank, G. Rock

JOANNEUM RESEARCH,
Institute for Water Resources Management –
Hydrogeology and Geophysics,
Elisabethstraße 16 / II
A-8010 Graz – Austria

## Abstract

In September 2001 a tracing experiment has been worked out at the groundwater test field "Wagna" (400 * 300 m in the Mur valley aquifer system) using 75 kg of sodium bromide as tracer and injected during 20 minutes at the groundwater table (pulse injection). Due to the density of the solute the tracer has been distributed over the whole aquifer depth in a very short time.

For estimating the longitudinal dispersivity an analytical model for calibrating Peclet' Typecurves on measured tracer breakthrough curves has been used. Transversal dispersivity was estimated using 2D-analytical model calibration. The results of parameter estimation on the field scale are compared to the results of a 2D-numerical bromide transport model for the test field based on a transient groundwater flow model.

## Keywords

Tracing Experiment; Parameter Estimation; Modelling

## Introduction

In 2001 at the research test field "Wagna" a new type of groundwater sampling site has been implemented for monitoring 3D ground water quality in shallow phreatic aquifers (Berg, 2003). The agricultural research area with a scale of 400 * 300 m is situated in the Murtal River Aquifer system in the southern part of Austria. In the saturated zone of the aquifer with a thickness ob about 3 m 36 sampling sites has been installed (Fig. 1), where samples can be taken in different depths and the hydraulic head was measured in short time intervals. From the cores of the bore holes the boundary between tertiary and quaternary sediments as aquiclude has been estimated. Groundwater recharge from infiltrating precipitation was measured at a lysimeter nearby.

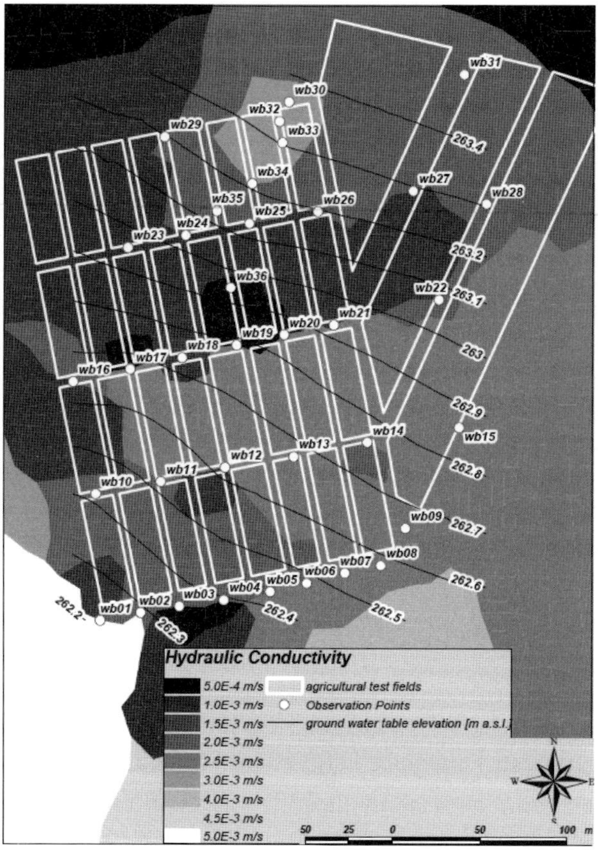


**Fig. 1.** Groundwater test field "Wagna": distribution of hydraulic conductivity and observation wells for 3D groundwater quality monitoring.

## Tracing experiment

In September 2001 a tracing experiment has been worked out using 75 kg of Sodium bromide as tracer diluted in 300 l of local groundwater and injected during 20 minutes at bore hole "wb30" (location see Fig. 1). The tracer solution has been injected at piezometer No. 3 10 cm below the actual groundwater table elevation (Fig. 2). The hydraulic reaction of the groundwater table on input has been observed at piezometer No. 4 and has been noted to be lower than 0.005 m. Due to the duration a pulse injection can be assumed. Vertical tracer distribution at injection place and tracer dilution was measured at piezometers No. 4 and No. 5. As visible in Fig. 2 the tracer concentration at injection point was lower than 1 ppm a week after injection.

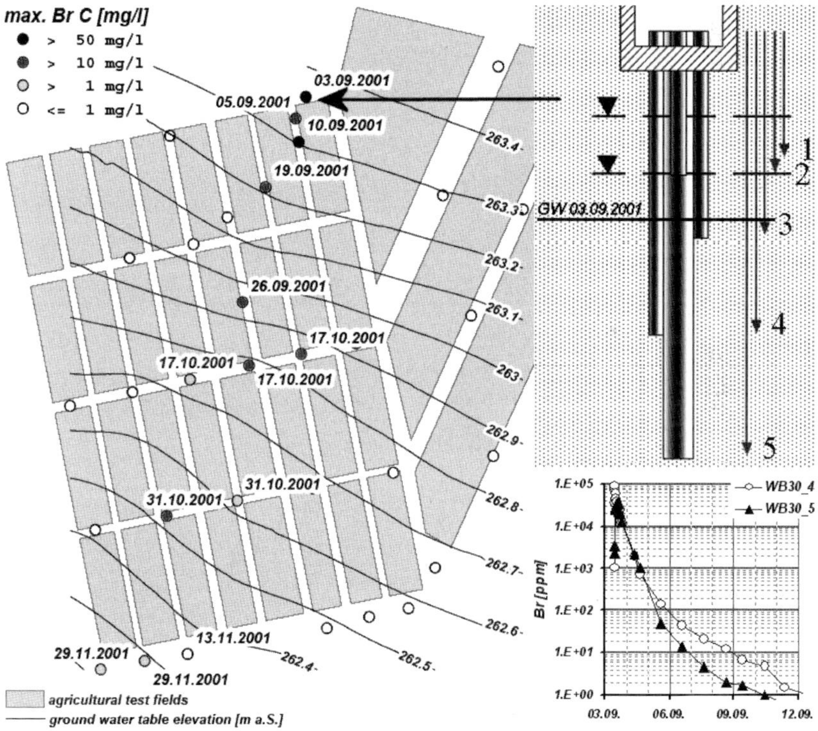


**Fig. 2.** Tracer (Br) injection at piezometer WB30_3 (September, 3rd 2001) 10 cm below ground water table. Control of tracer distribution with depth and dilution in piezometer WB30_4 and WB30_5 (concentration / time – diagram). Arrival time of tracer peak and peak concentration in the flow field downstream injection piezometer.

In Fig. 2 the distribution of the tracer cloud at the groundwater test field is shown as well, showing the arrival time of the tracer concentration peak and the measured peak concentration at the different observation wells. As expected the measured data indicate a movement of the tracer in ground water flow direction and a decrease of tracer concentration due to dilution and dispersion effects.

Due to aquifer geometry we assume 2D-groundwater flow and transport in horizontal direction. The homogeneity of vertical tracer distribution is visible in Fig. 3, where measured tracer concentration over time in different depths in the centre of the test field (“wb19”, location see Fig. 1) are illustrated. In correlation to aquifer depth WB19_3 is situated close to the groundwater table, WB19_5 close to the aquiclude and WB19_4 in the middle between them in the center of the aquifer.

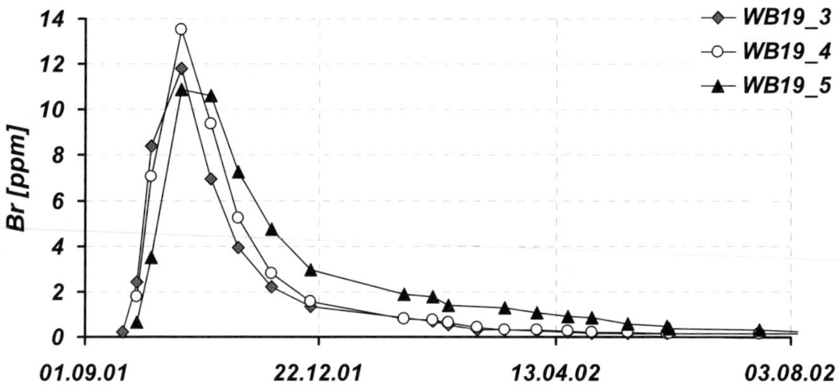


**Fig. 3.** Tracer (Br) distribution with depth in piezometers WB19_3, WB19_4 and WB19_5 in the center of the flow field (location see Fig. 1)

## Parameter Estimation

The tracer breakthrough at the different measuring points has been used to estimate mean groundwater flow velocity ($v_a$) from the center of gravity of tracer concentration over time. The local hydraulic gradient (I) was estimated from hydraulic head measurement and interpolated contour lines of groundwater table elevation. From lysimeter investigations at the test field and evaluating groundwater table fluctuation (Fank, 1999) we assume a drain- and fillable pore volume ($n_e$) of about 0.15. Darcy’s law led us to the estimation of hydraulic conductivity: $k_f = (v_a * n_e) / I$. The resulting set of hydraulic parameters for the different observation wells is shown in Table 1.

**Table 1.** Estimation of hydraulic and tracer transport parameters resulting from tracer test evaluation with: D = Distance between injection point and observation point, $v_a$ = groundwater flow velocity, I = local hydraulic gradient, $\alpha_L$ = longitudinal dispersivity, $\alpha_T$ = transversal dispersivity, $k_f$ = hydraulic conductivity (location of observation points see Fig. 1)

| observation point | D [m] | $v_a$ [m/d] | I [ ] | $\alpha_L$ [m] | $\alpha_T$ [m] | $\alpha_L/\alpha_T$ | $k_f$ [m/s] | $\alpha_L$/D [ ] | $\alpha_T$/D [ ] |
|---|---|---|---|---|---|---|---|---|---|
| WB01 | 312 | 3.00 | 0.004 | 18.35 | 1.08 | 17.00 | 0.002 | 0.059 | 0.003 |
| WB03 | 275 | 4.52 | 0.003 | 19.64 | 1.03 | 19.00 | 0.003 | 0.071 | 0.004 |
| WB11 | 227 | 3.23 | 0.003 | 12.61 | 0.63 | 20.00 | 0.002 | 0.056 | 0.003 |
| WB12 | 220 | 3.69 | 0.003 | 14.67 | 0.73 | 20.00 | 0.002 | 0.067 | 0.003 |
| WB18 | 155 | 4.44 | 0.004 | 10.33 | 0.34 | 30.00 | 0.002 | 0.067 | 0.002 |
| WB19 | 146 | 2.92 | 0.004 | 9.13 | 0.37 | 25.00 | 0.001 | 0.063 | 0.003 |
| WB20 | 135 | 3.55 | 0.004 | 9.00 | 0.36 | 25.00 | 0.002 | 0.067 | 0.003 |
| WB36 | 110 | 4.02 | 0.003 | 8.46 | 0.42 | 20.00 | 0.002 | 0.077 | 0.004 |
| WB33 | 23 | 2.76 | 0.001 | 4.66 | 0.17 | 28.00 | 0.006 | 0.200 | 0.007 |
| WB34 | 51 | 4.70 | 0.003 | 5.67 | 0.27 | 21.00 | 0.004 | 0.111 | 0.005 |

The estimated spatially distributed hydraulic conductivity was used as initial value for the calibration of a transient groundwater flow model for the test field. The boundary conditions have been taken as time series of potential head from a regional groundwater flow model. The calibrated distribution of hydraulic conductivity is shown in Fig. 1. Fig. 4 demonstrates the quality of calibration comparing measured and calculated hydraulic head at different observation points (location see Fig. 1).

The horizontal black line in Fig. 4 indicates the duration of tracing experiment at the test field. We assume the groundwater flow direction and velocity as well described by the groundwater flow model for the whole time scale of the tracing experiment.

To describe the spreading of solutes from the flow paths in saturated groundwater flow the concept of dispersion is mostly used. To estimate the coefficient of dispersion from tracing experiments the calculation of "Peclet typecurves" may be used (Sauty 1977). For one-dimensional tracer transport from injection point (wb30 in Fig. 1) in groundwater flow direction normalized time – concentration curves were calculated and fitted to measured tracer breakthrough curves varying the Peclet number (Fig. 5). Using the fitted Peclet number longitudinal dispersion coefficient is calculated using the equation: $D_L = (v_a * D) / Pe$, where $D_L$ = longitudinal dispersion coefficient, $v_a$ = flow velocity, D = flow Distance (Table 1) and Pe = Peclet number. Longitudinal dispersivity ($\alpha_L$) is estimated from ($D_L/v_a$).

Assuming the tracer application at wb30 a pulse injection comparing the injection time to the tracer velocity and assuming 2D parallel groundwater flow in an homogeneous and isotropic porous medium the partial differential advection-dispersion transport equation has an analytical solution. Using the analytical model of Schulz (1992) transversal dispersivity ($\alpha_T$) may be calibrated on measured tracer distribution over time (Fig. 6 shows calibration quality on WB19_4 and resulting $\alpha_T$ for example). Calibrated values for $\alpha_T$ on all observation points which are tangent to the tracer cloud are shown in Table 1.

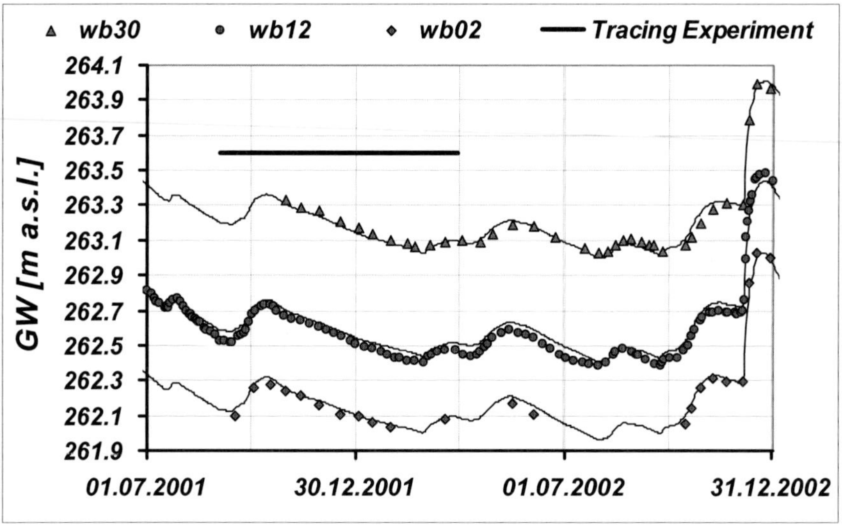


**Fig. 4.** Calibration of groundwater flow model – comparison of measured (points) and calculated (line) hydraulic head at different observation points (location see Fig. 1) and duration of tracing experiment (indicated with the horizontal black line).

| | | | |
|---|---|---|---|
| **PECLET-Number** | ***Pe*** | | **15** |
| **Distance Injection Point - Observation Point [m]** | ***x*** | | **146** |
| **mean groundwater velocity (tracing experiment) [m/s]** | $v_a$ | **C - center of gravity** | **3.0826E-05** |
| **longitudinal dispersion coeffizient [m²/s]** | $D_l$ | $(v_a * x) / Pe$ | **0.00030004** |
| **longitudinal dispersivity [m]** | $\alpha$ | $(D_l / v_a)$ | **9.73** |

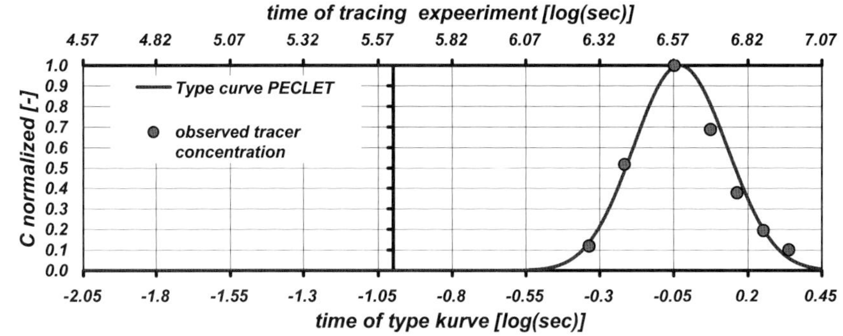


**Fig. 5.** Calibration of the Peclet's type curve evaluation (Sauty, 1977) at observation point WB19_4 (location see Fig. 1) and calculated transport parameters in groundwater flow direction using the analytical model.

| | | | |
|---|---|---|---|
| **tracer mass [kg]** | $\Delta M$ | Br | 58.243 |
| **aquifer thickness [m]** | m | borehole logging | 3 |
| **effective porosity [-]** | $n_f$ | evaluation of groundwater hydrograph | 0.17 |
| **longitudinal dispersivity [m]** | $\alpha_L$ | PECLET - evaluation | 9.73 |
| **transversal dispersivity [m]** | $\alpha_T$ | Kalibration | 0.61 |
| **flow distance x - direction [m]** | x | flow net | 146.00 |
| **flow distance y - direction [m]** | y | flow net | 15.00 |
| **mean groundwater flow velocity [m/d]** | $v_a$ | Tracer breakthrough - center of gravity | 3.25 |

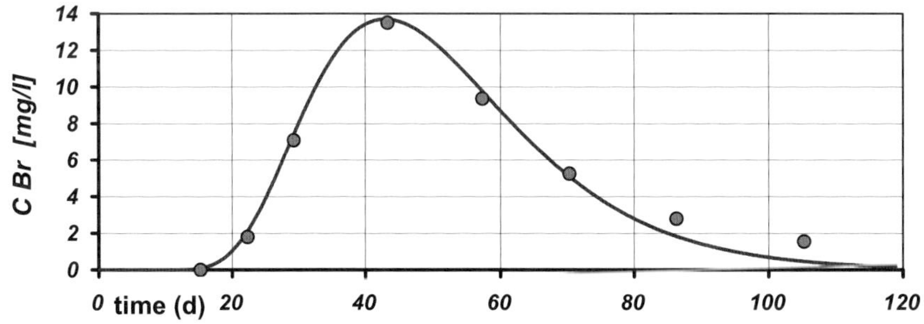


**Fig. 6.** Calibration of a 2D groundwater transport model (Schulz 1992) at observation point WB19_4 (location of "wb19" see Fig. 1) and calculated groundwater transport (dispersivity) parameter perpendicular to flow direction using the analytical model.

Schulz (1992) remarks a wide variety for longitudinal dispersivity in sandy and gravely porous aquifer systems. In field experiments values between 0.1 and 100 m for $\alpha_L$ have been investigated. Transversal dispersivity in most cases is much lower than $\alpha_L$. A very wide variety of the relation $\alpha_L$:$\alpha_T$ between 2:1 and 400:1 is given. For most cases a relation of 5:1 to 10:1 is assumed. In our experiment values of $\alpha_L$ in the range between 5 to 20 m were calculated with a relation of $\alpha_L$:$\alpha_T$ from 17:1 to 30:1. The values shown in Table 1 agree very well with literature data.

A dependency of $\alpha_L$ on flow length is stated as well (Schulz, 1992). The same effect has been observed at the tracing experiment in Wagna. The dependency of $\alpha_L$ on the distance of the observation point from the injection point is shown in Fig. 7. At the observed distances between 25 and 300 m a linear correlation with high significance was found for $\alpha_L$. A similar correlation has been investigated for $\alpha_T$ (Fig. 7). The concept of dispersivity assumes $\alpha$ to describe the in-homogeneity of the aquifer and be sediment constant. In most numerical models dispersivity is implemented as a material constant. Looking on literature data (Schulz, 1992) and on the results of the presented experiment a parameter like $\alpha$/D (Table 1) describing solute distribution as a material parameter will give better information for the interpretation of tracer distribution from point source at field scale.

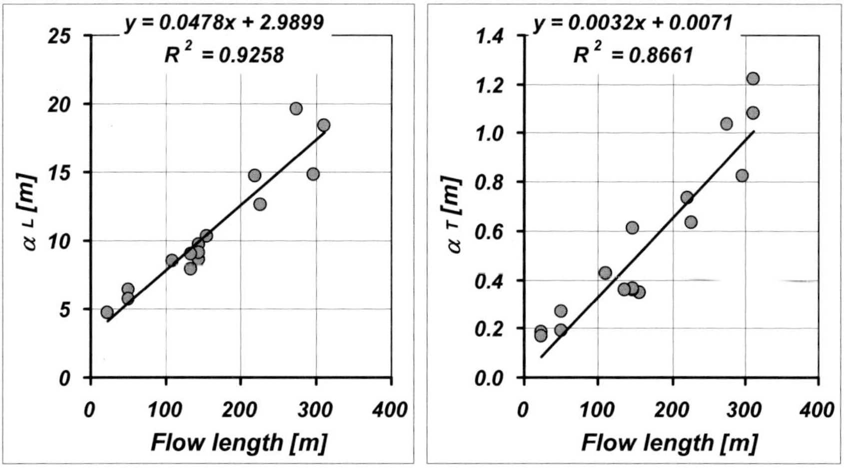


**Fig. 7.** Estimation of longitudinal and transversal dispersivity from tracing experiment using analytical modeling.

## Calibration of a numerical tracer transport model

The results of analytical modelling to determine $\alpha_L$ and $\alpha_T$ were used as initial conditions for the calibration of a 2D horizontal numerical tracer transport model. Measured tracer breakthrough was used for calibration of the model. As shown in Fig. 8 the calibration of the model is well done on the central flow line of tracer transport with only few modification of longitudinal dispersivity: the spatial distribution of $\alpha_L$ show values between 2 m near injection point and 20 m at the end of the test field in a flow distance of about 300 m downstream "wb30".

Problems arise in calibrating the transport model on observation points beside the central flow line as shown in Fig. 9: Although the results of analytical modeling $\alpha_T$ have been modified in some regions to values higher than 3 m it was not possible to fit measured tracer breakthrough and numerical modeling results at observation points "wb02" and "wb03". Calculated tracer concentration values are much higher than the measured ones.

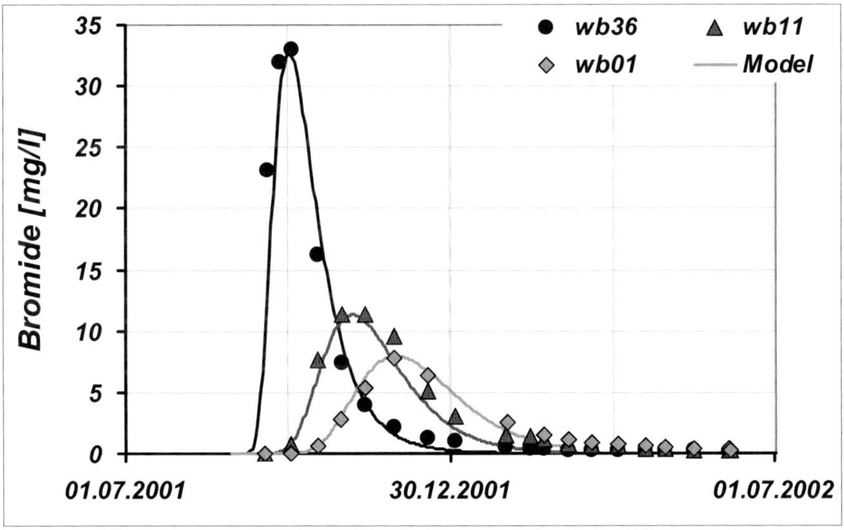


**Fig. 8.** Measured Bromide concentration breakthrough curves compared to the results of numerical modeling at observation points in central flow direction (location of observation points see Fig. 1)

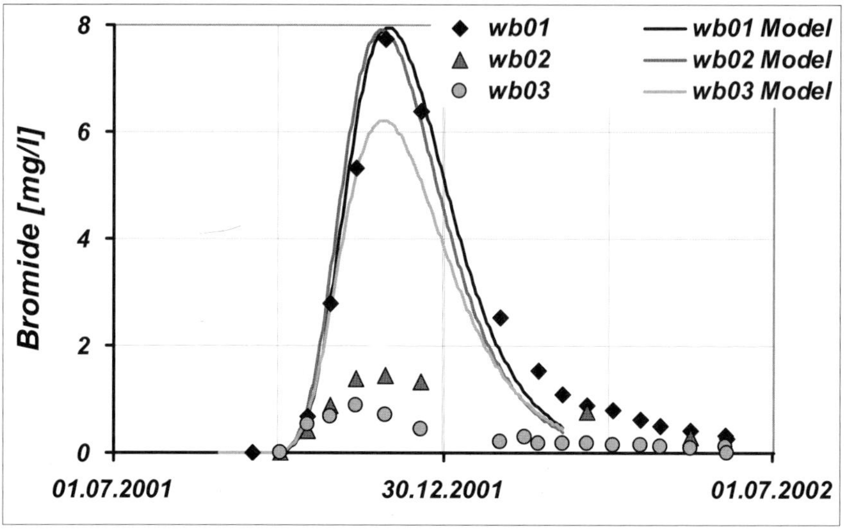


**Fig. 9.** Measured Bromide concentration breakthrough curves compared to the results of numerical modeling at observation points perpendicular to flow direction (location of observation points see Fig. 1)

## Conclusion

The evaluation of a tracing experiment and the attempts to calculate tracer breakthrough using a 2D horizontal Finite Element transport model at different observation points in an intensively monitored groundwater test field in a shallow phreatic aquifer system led us to some conclusions:

- Monitoring tracer concentration data gives the possibility to visualize transport processes in groundwater
- The results of analytical modeling to estimate longitudinal dispersivity are well comparable to calibrated values used in a 2D horizontal Finite Element transport model
- Using the values of transversal dispersivity gathered from analytical modeling for calibrating a numerical transport model gives an overestimation of tracer concentration at observation points beside the central groundwater flow line in the test field
- The transversal dispersivity parameter in numerical modeling is extremely sensitive to unobserved transient groundwater flow processes.

## References

Berg W (2003) Monitoring, Analysis and Interpretation of Nitrogenous components and their layered transport into the groundwater of a shallow quarternary aquifer (Leibnitzer Feld, WAGNA). PHD-Thesis, Inst f Geographie und Raumforschung, Karl Franzens Universität Graz, p 157

Fank J (1999) Die Bedeutung der ungesättigten Zone für Grundwasserneubildung und Nitratbefrachtung des Grundwassers in quartären Lockersediment-Aquiferen am Beispiel des Leibnitzer Feldes (Steiermark, Österreich). Beiträge zur Hydrogeologie, 49/50, pp. 101-388

Schulz HD (1992) Physikalische Grundlagen des Stofftransports im Untergrund. Geohydrologische Markierungstechnik, Lehrbuch der Hydrogeologie, 9, pp. 325-362,(Borntraeger) Stuttgart

Sauty JP (1977) Contribution à l'identification des paramètres de dispersion dans les aquifères par interprétation des expériences de tracage. – Bur. de Recherches Géol. et Minières; Dépt Hydrogéol 77SGN515HYD, p 157 , Orléans

# Biogeochemical modeling of reactive transport applied to laboratory and field studies on jet-fuel contamination

P. Aagaard[1], J.B.S. Knudsen[2], M.R. Klonowski[3], G. Breedveld[1,4], Z. Zheng[5]

[1] Department of Geosciences, University of Oslo, Norway
[2] Norconsult, Sandvika, Norway
[3] Polish Geological Institute, Wroclaw, Poland
[4] Norwegian Geotechnical Institute, Oslo, Norway
[5] Lawrence Berkeley National Laboratory, Berkeley, California, USA

Corresponding author: P.O.Box 1047 Blindern, 0316 Oslo – Norway
Tel: +47.22856644 – fax: +47.22854215 – e-mail: per.aagaard@geo.uio.no

## Abstract

Within the PHREEQC framework, a dual Monod kinetics formulation has been included, which allows rate dependencies of both substrates, terminal electron acceptors and inhibitors. In this way, PHREEQC will simulate the redox processes under concern both with regard to kinetics and thermodynamics. Furthermore, PHREEQC allows one-dimensional reactive transport to be simulated.

The biogeochemical processes involved in transport and biodegradation of dissolved jet-fuel were simulated for two cases by PHREEQC with this Monod kinetics scheme. The column studies of Knudsen (2003) exploring dissolved jet-fuel transport and biodegradation within initially pristine aquifer sediments dominated by pyrite oxidation and calcite dissolution. Pyrite oxidation will compete with aerobic biodegradation, thereby reducing the efficiency of aerobic bioremediation. The 1D column simulations gave reasonable agreement with measured biodegradation, mineralization and pyrite oxidation rate, and reproduced the overall microbial processes well, but they failed to mimic the observed ferrous iron. A dual porosity approach should be included.

The second case with a jet-fuel contaminated plume under monitoring was simulated with a 1D PHREEQC column from a plume cross section along the flow direction. The biogeochemical reactions themselves were described reasonably well, but dispersional/diffusional transport effects could not be simulated sufficiently with a 1D column of PHREEQC alone. Here truly coupled models of 3D flow and biogeochemical reactions must be applied.

## Keywords

Groundwater contamination; jet-fuel; bioremediation; modeling

## Introduction

Laboratory and field studies on biodegradation and transport of de-icing chemicals and the water soluble part of jet-fuel with relevance to aviation activity on the Gardermoen aquifer in southeastern Norway, have accumulated a large data-set during the last ten years (Aagaard and Tuttle 1996). This research has been applied in subsurface treatment plants used in the daily operation of Oslo International Airport, located directly above the Gardermoen phreatic aquifer (Breedveld et al. 1997, 1999).

Various modeling efforts have been taken in the analysis of the observational data (Kitterød and Finsterle 2000, French et al. 2001, Søvik et al. 2002, Alfnes et al. 2003 a,b, Wong 2003). However, in order to integrate microbial degradation kinetics with abiotic processes in groundwater, we have found the PHREEQC model to be quite useful in both the process interpretation and transport predictions (Aagaard et al. 2001, Zheng et al. 2002, Knudsen 2003). There is a large variation in scale of the studies, the soils have been both pristine and previously contaminated, substrate concentrations have varied considerably, and microbial degradation has taken place under changing aerobic and anaerobic conditions. The aim of this communication is give some examples of biogeochemical simulations by PHREEQC to demonstrate its applicability.

The scenario of concern is pollution of dissolved jet-fuel in the Gardermoen aquifer. Two cases are considered: one relates to biodegradation under aerobic/anaerobic conditions created by hypothetical minor spills of jet-fuel, the second to anaerobic biodegradation of methylated benzenes and naphthalene in a plume where natural attenuation is presently monitored.

## The model

Within the PHREEQC framework, a dual Monod formulation (Essaid et al. 1995) has been included, which allows rate dependencies of substrates, terminal electron acceptors and inhibitors. We have applied maximum microbial density to limit growth of aerobic bacteria, and the inhibition/switch functions by the terminal electron acceptors. In this way, PHREEQC will simulate the redox processes under concern both with regard to kinetics of biodegradation and the partial equilibrium conditions imposed by PHREEQC, as the code calculates the simultaneous consequences of aqueous complexation, mineral equilibrium and surface complexation/ion exchange. Additional kinetic formulations of mineral dissolution/precipitation may also be added. And not least, PHREEQC allows one-dimensional reactive transport to be simulated.

The growth of bacteria from the mineralization of organic carbon was described by Monod kinetics. Its general expression is shown for one arbitrary microbial group:

$$\left(\frac{\partial X_r}{\partial t}\right)_{growth} = \left(\mu_{\max,r} \cdot X_r \cdot Y_r \cdot \prod_{m=1}^{nrmonod} \frac{C_m}{K_m + C_m} \cdot \prod_{i=1}^{nrinhib} \frac{dinhib_i}{dinhib_i + C_i}\right) - \left(\mu_{dec,r} \cdot X_r\right) \qquad (1)$$

$X_r$ is the concentration of the microbial species that is growing or decaying by the respirative pathway r. $\mu_{max,r}$, $\mu_{dec,\,r}$, and $Y_r$ are the corresponding maximum specific growth rate, constant decay rate ($1 \cdot day^{-1}$) and yield coefficient (mol org. $C_{cell}$/mol org. $C_{degraded}$). *nrmonod* and *nrinhib* are the Monod growth and inhibition terms that affect microbial growth. $C_m$ and $C_i$ are the concentration of species (substrate and terminal electron acceptors) that affect microbial growth or inhibition. $K_m$ and $dinhib_i$ are the corresponding half saturation and inhibition constant ($mmol \cdot L^{-1}$). The half saturation constant is defined as the substrate concentration at which the microbial growth rate $\partial X/\partial t$ is half of the maximum growth rate ($\mu_{max,}$). It indicates the affinity of the microbial population for the substrate. Thus, increasing half saturation decreases the substrate utilization and growth rate. For each respirative pathway microbial growth depends on two Monod terms. One is the organic carbon substrate term, and the other the electron acceptor (either $O_2$, Fe(III) or $SO_4^{2-}$) term. Inhibition terms of the type shown above were first developed by Widdowson et al. (1988). They are used to approximate inhibition processes such as the suppression of strictly anaerobic bacteria by dissolved oxygen (DO), or the unlimited growth of bacteria and consequent clogging of the column (Kindred and Celia 1989). In this experiment the maximum microbial density limited the growth of aerobic bacteria at the column inlet, whereas the presence of DO limited the growth of anaerobic bacteria (i.e. iron and sulfate reducers).

From the changes in biomass concentrations according to (1), the changes in organic carbon substrate concentration were calculated as follows,

$$-\frac{\partial C_{OC}}{\partial t} = \sum_{r=1}^{nrr} X_r \cdot \mu_{\max,r} \left( \prod_{m=1}^{nrmonod} \frac{C_m}{K_m + C_m} \right) \tag{2}$$

Where $C_{OC}$ is the organic substrate concentration and the remaining symbols ($X_r$, $\mu_{max,\, r}$, $C_m$ and $K_m$) are defined from the preceding discussion. The consumption of electron acceptors and the release of reduced species are calculated internally from the defined reaction stoichiometry.

Microbial populations responsible for biodegradation of the hydrocarbons were assumed to be heterotrophic bacteria with a uniform initial distribution. The microbial activity was approximated by the bacterial biomass. The Monod equations assume biodegradation to start immediately without any lag period. Lack of organic substrate and the terminal electron acceptor were assumed to control microbial metabolism, as nutrient limitations due to lack of nitrate or phosphate were not assumed to occur. We also assumed complete mineralization of the organic compounds to $CO_2$ and water. All physical parameters were assumed to be constant during the calculations.

## Parameter estimations

The Monod formulation offers advantages over traditional 0. or 1. order degradation kinetics, as it can be applied to the wide range of substrate concentrations encountered in organic contaminant plumes. However, it also requires a large set of model parameters to be estimated. This is often not straight forward, and it may not be be possible to obtain unique parameter sets based on limited experimental data. We have adopted literature data (Essaid et al. 1995), and parameter constraints (Schäfer et al. 1998) to limit the numbers of parameters to be estimated.

## Simulations

*Case 1: Biodegradation under aerobic/anaerobic conditions by hypothetical minor spills of jet-fuel.*

The scenario is dissolved jet-fuel transport and biodegradation within initially pristine aquifer sediments dominated by pyrite oxidation and calcite dissolution. Pyrite oxidation will compete with aerobic biodegradation, thereby reducing the efficiency of remediation schemes which rely on aerobic bioremediation (Thullner and Schäfer 1999). The observational basis for the simulation is a laboratory column experiment (Knudsen 2003) where a mixture of methylated benzenes and naphthalene was passed trough a column reactor consisting of sandy aquifer material. The pristine aquifer sand was collected as an undisturbed drill core at 8-9 meters depth on the outer part of the Gardermoen ice-contact delta (Tuttle 1997). Influent and effluent concentrations of organic and inorganic species were monitored at fixed time intervals for 40 days.

A cumulative carbon balance (Figure 1), based on influent/effluent chemistry, shows that of the total of 40 mmol organic carbon loaded into the column, 7 mmol was sorbed, 4 mmol (10%) passed through without being affected by biodegradation, 14 mmol was completely mineralized, while 3 mmol turned into biomass and the rest 13 mmol were organic intermediates. A corresponding mass balance of dissolved inorganic carbon (DIC), shows that a sulfate reduction contribution is questionable, but cannot totally be excluded. The contribution to DIC from aerobic biodegradation, omitting sulfate reduction, corresponds to a 90%. If sulfate reduction is included, anaerobic microbial processes account for a larger part, approximately 40%.

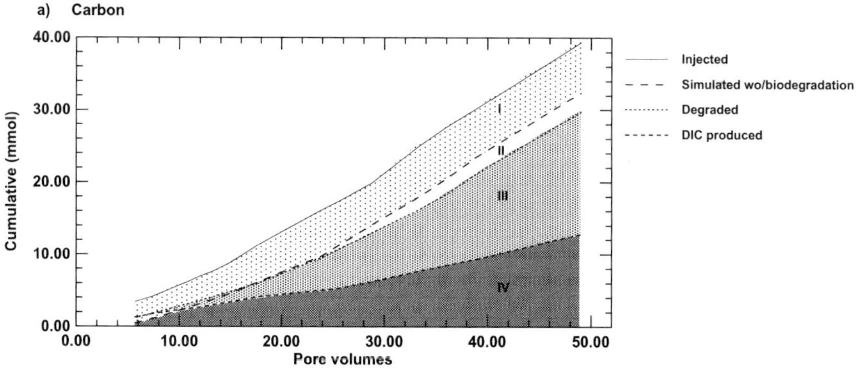


**Fig. 1.** Cumulative carbon balance based on influent/effluent chemistry. Of the total load of 40 mmol organic carbon, 3 mmol passed through the column, 7 mmol was sorbed and the rest partially or completely removed by biodegradation. The complete mineralization to dissolved inorganic carbon (DIC) accounted for 14 mmol.

The PHREEQC computer code (Parkhurst and Appelo 1999) was used with cation exchange and anion surface adsorption processes included. Specific kinetic formulations of pyrite oxidation and calcite dissolution/precipitation were written based on Williamson and Rimstidt (1994) and Plummer et al. (1978). Other mineral water reactions occuring are thus modeled as equilibrium reactions, with mass-transfer caused by other reactions. The organic solute, i.e. the mixture of toluene, o-xylene, 1,2,4 trimethyl benzene and naphthalene, was lumped into one weighted organic carbon compound with 8.30 carbons and 9.74 hydrogen respectively. This was justified, based on a biodegradation kinetics in the column reactor that could be described with a common first order kinetics per organic carbon (Knudsen 2003). Retardation of organic solute was described by linear instantaneous sorption with an average $K_d$ of 1.5. Biodegradation of the organic substrate was expressed by Monod kinetics with appropriate growth, inhibition and decay terms for the microbial population. In our case, the maximum microbial density limited the growth of aerobic bacteria at the column inlet, whereas the presence of dissolved oxygen limited the growth of anaerobic bacteria (i.e. iron and sulfate reducers).
A low value of initial biomass of 1·10-6 mol·$L^{-1}$ was used, similar to other reactive transport simulations (Schäfer et al. 1998, Zheng et al. 2002). We assigned yield coefficients according to the method of Schäfer et al. (1998). The remaining parameters were fitted with PEST, while the $K_m$ values for TEAP's were taken from Schäfer et al. (1998). To maintain a similar specific turnover rate for all three microbial groups, the ratio of the maximum growth rate and yield coefficient was held constant. This also simplifies parameter estimation since only one maximum growth rate ($\mu_{max}$) needs to be optimized with PEST. The decay rate ($\mu_{dec}$) was set equal to ($\mu_{max}$/100), which is consistent with that used by Schäfer et al. (1998). In addition, initial amounts of reactive pyrite, Fe(III)-hydroxide and calcite, as well as surface to volume (A/V) ratio of pyrite and calcite, were also estimated by PEST. To enable the growth of iron reducing bacteria with the consequent release of Fe(II) it was necessary to specify a rather high supersaturation for ferrihydrate (i.e. a saturation index, SI: 3). This was justified based on the fact that bacteria have been observed to use extracellular complexing agents to increase the solubility of the solid Fe(III) in the sediment (Lovley et al. 1994).

The contribution to dissolved inorganic carbon (DIC) from aerobic biodegradation corresponded to 80-90%, with iron reduction dominating the anaerobic degradation. This was also evident in the simulations where the temporal evolution in solid and bacterial species is depicted in Figures 2a. The transient changes in the vicinity of the column outlet, there is a slight increase in Fe(OH)3(a) precipitated from pyrite oxidation, but as conditions become anaerobic and the aerobic bacteria start to decay at about 14 pore volumes, Fe(OH)3(a) decreases rapidly. At the same time the Fe(III)-reducers grow and reach maximum density at 34 pore volumes, after which point they decay due to lack of Fe(OH)3(a). The increase in sulfate reducers is very small, and is nearly negligible compared to the aerobes and Fe(III)-reducers.

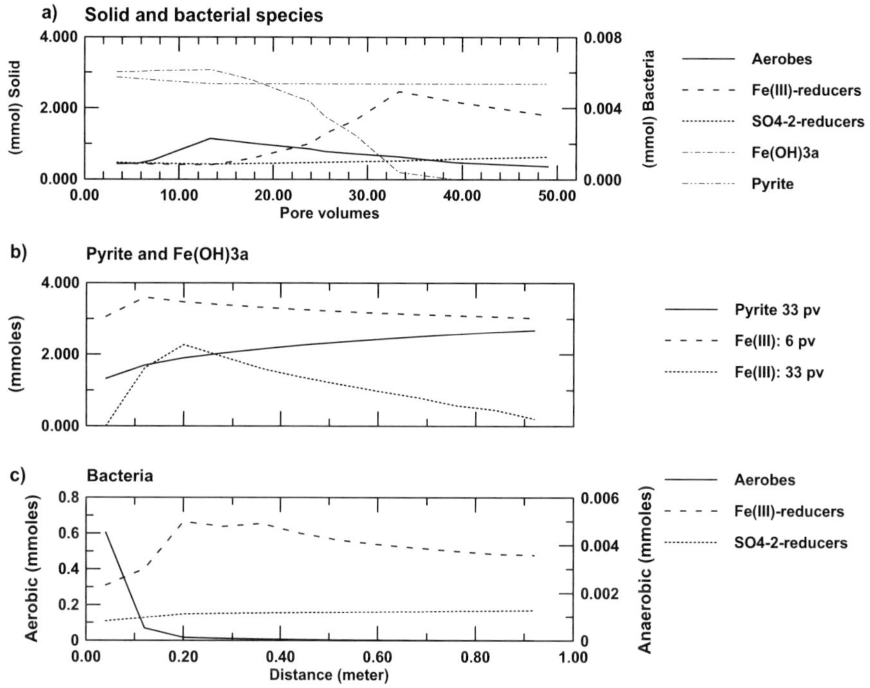


**Fig. 2.** Simulated profiles of the evolution in solid and bacterial species with time at the column effluent in (a) and along the length of the column at 6 and 33 pore volumes in (b) and at the end of column operation in (c). X-axis in (b) and (c) denotes distance from column inlet.

The spatial distribution is illustrated in Figure 2b,c. Along the length of the column in Figure 2c, the aerobes initially dominate. The aerobic bacteria reach a maximum density of 0.9 mmol·kg-1 organic cell carbon at the inlet and quickly decay, as dissolved oxygen decreases, at which point the Fe(III)- and sulfate reducers start to grow. Due to their lower yield coefficient ($Y_i$), their density is significantly smaller than the aerobes. The Fe(III)-reducers increase near the inlet due to the added amount of Fe(OH)3(a) from pyrite oxidation.

Aerobic processes clearly dominate close to the inlet (the first 10 cm of the column); this is also where the major part of the biodegradation takes place. There is initially a rapid pyrite oxidation in this part of the column, However, with time the bacterial biomass builds up, causing the aerobic biodegradation nearly to out-compete pyrite oxidation. The anaerobic biodegradation is dominated by Fe(III) reduction. However, as reactive Fe(III) hydroxide decreases, some sulfate reduction, also appears to take place towards the column outlet.

The simulations demonstrated reasonable agreement with measured biodegradation, mineralization and pyrite oxidation rate, and reproduced the overall microbial processes well, but they failed to mimic the observed ferrous iron. In the experiments, the release of ferrous iron starts early before the depletion of dissolved oxygen in the effluent and in concentrations greatly exceeding that expected from pyrite oxidation. This demonstrates that aerobic pyrite oxidation occurs at the same time as reductive dissolution of ferric hydroxide. The calculated $Fe^{2+}$ release, however, is retarded until pyrite oxidation ceases. This strongly indicates simultaneous aerobic/anaerobic reactions in adjacent micro-domains in the column. Our initially homogeneous column model is thus not able to simulate such apparent heterogeneity in physical/ chemical/ microbial properties. This would call for dual porosity approaches (Van Genuchten 1985).

*Case 2 Anaerobic biodegradation of methylated benzenes*

The study site is situated at the fire fighting training facility of the Gardermoen airport, about 60 km NE of Oslo, SE Norway (Fig 1). The assumed source of pollution, (a leaking oil skimmer) was removed in autumn 1998, and simultaneously 17.5 litres of hydrocarbons as free phase and about 6000 litres of polluted groundwater was recovered and treated (Aagaard et al. 2001). A network of piezo-meters was installed to monitor the groundwater movement. Based on initial observation of groundwater chemistry, remediation using natural degradation of the groundwater contaminants was suggested (Rudolph-Lund and Sparrevik, 1999). This was approved by the Norwegian State Pollution Control Agency as a pilot study on natural attenuation. To be able to document the fate and transport of the contaminants in the plume an extensive network of multilevel monitoring points was established (Klonowski et al. 2002, Klonowski 2005). An elongated plume of residual free phase jet-fuel of m is associated with the groundwater table, and seems to have been divided vertically into mobile and immobile zones because of vertical smearing due to groundwater table fluctuations. Directly below the residual free phase, there is a pollution plume of the dissolved aromatic fraction and associated inorganic zonations. The horizontal extent of this plume has remained rather stable during a 3 years monitoring (Klonowski 2005), indicating that a strong intrinsic bioremediation has taken place.

Among the dissolved hydrocarbons, toluene is only found centrally in the plume at very low concentrations, indicating a rapid biodegradation. Naphthalene is the most recalcitrant of all the jet fuel components followed by trimethylbenzenes, whereas m/p-xylene, ethyl benzene and o-xylene showed a distribution which indicated an increasing biodegradation from m/p-xylene to ethyl benzene and finally o-xylene. A TEAP mass balance of the plume indicates that both aerobic and anaerobic biodegradation takes place, with sulfate reduction and iron reduction to be dominant in the core of the plume.

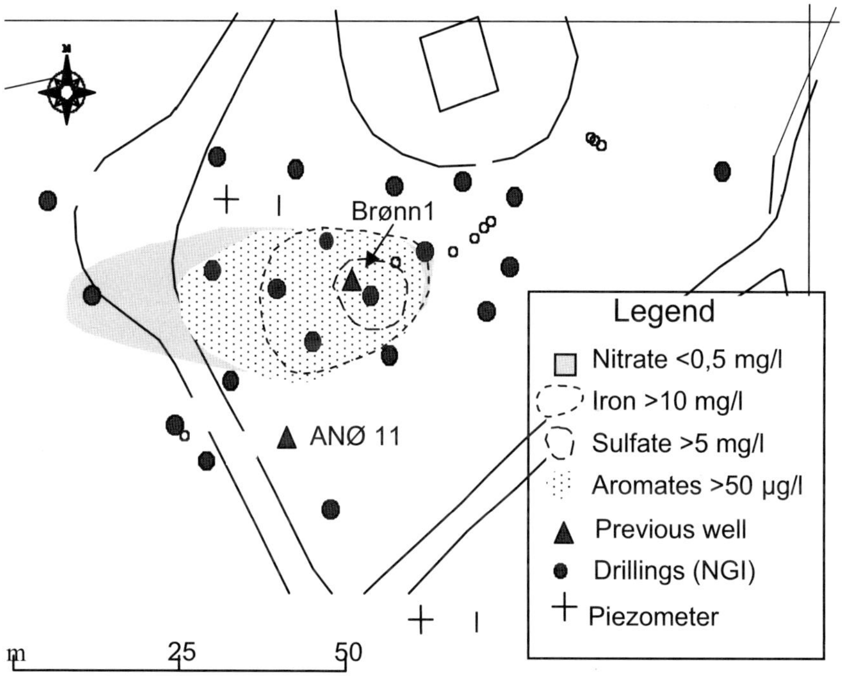


**Fig. 3.** Initial delination of the pollution plume at fire-fighter training ground at Gardermoen, mapped by drillings and observation of groundwater chemistry (Rudolph-Lund and Sparrevik, 1999).

Zheng et al. (2002) performed laboratory batch and column experiments with contaminated aquifer sediments, local groundwater and four soluble aromatic components in order to study biodegradation under anaerobic conditions. Terminal electron acceptor processes (TEAPs) observed were iron reduction, sulfate reduction and, at later stage of incubation and/or electron acceptor consumption, also methanogenesis. Monod parameters used to simulate these experiments were fitted from a base of those listed by Essaid et al. (1995), and were in a similar range for both batch and column condition (Zheng et al. 2002). Zheng also used the same parameter set to simulate field conditions at the fire-fighting training facility, he only reduced the initial biomass in the field case. In his simulations, Zheng used one common biomass to characterize the activity of all active bacteria groups and he had no upper boundary of biomass growth. We have thus adopted the approach of Knudsen (2003) with separate group biomass and asymptotical limits to bacterial growth, and used his rate parameters. They are consistent with those of Zheng for iron and sulfate reduction. We have also included the parameters for methanogenesis from Zheng (2002).

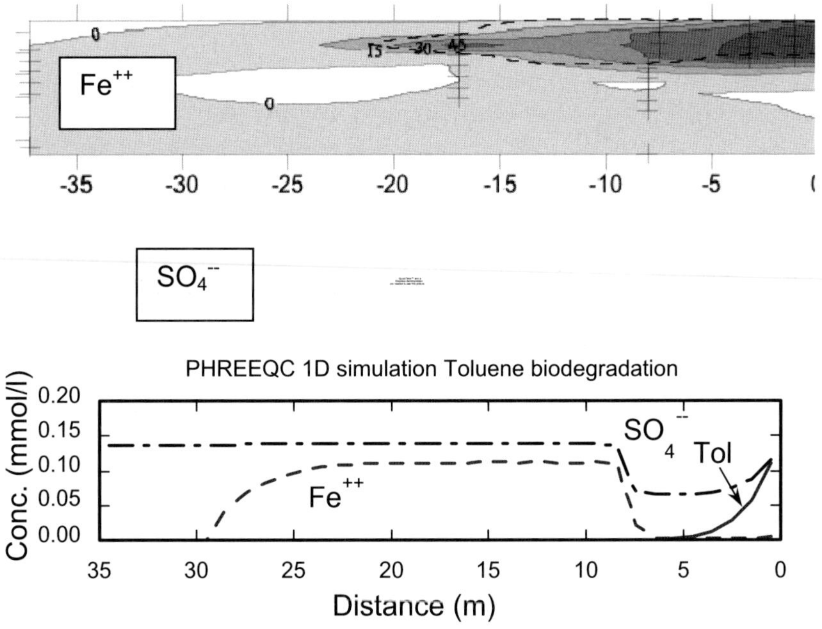


**Fig. 4.** Contaminant plume at the fire-fighter training ground at Gardermoen: Monitored (2003) profiles of ferrous iron and sulfate with dotted line showing the extent of the dissolved aromates (TDHC). Lower profile depicts result of 1D Phreeqc simulation of toluene transport and biodegradation from a constant source entering the field column from the right. Simulation period is 300 days. A stable dissolved toluene plume of 4-5 m length has been established within 30 days.

The following simplified 1D simulation has been run, with toluene as a single substrate. Toluene is here a model compound representing the rapidly degrading part of the alkylbenzenes. A 60 m long field column parallel with the groundwater flow direction is located from the plume center and outwards. Anaerobic toluene contaminated groundwater enters the column at the center, and displaces and reacts with the original pristine groundwater and sediments. Within a month a stable degradation pattern of toluene was established, where it was degraded within 4-5 meters downstream (see Fig. 4). This is in reasonable agreement with the field condition, knowing that toluene degrades rapidly and is only found in the plume core. The other alkylated benzenes are found further downstream, and with naphthalene as the most recalcitrant defining the outer boundary of the dissolved hydrocarbon plume. At the same time the simulation gave the expected plume development with an outer zone of ferrous iron ($Fe^{2+}$), and an inner zone of sulfate reduction and some methanogenesis. However, the width of these zones did not stabilize but continued to expand with time. Monitoring in the field has shown a rather stable plume size and zonation over a 3 years period (Klonowski 2005), with a small tendency to shrinkage. The main reasons for this discrepancy, appears to be the difference between a 1D column and a true 3D field situation. Under the column constraints, only longitudinal dispersion can take place, while in the field, the transverse dispersion and diffusion will greatly influence the concentration of electro-active dissolved species. In the column case, the influent contaminant solution will fill the column cross section, while in the real field condition the contaminant plume is only occupying a rather limited volume, thus increasing the transverse dispersion/diffusion transport. To include these effects of a 3D geometry, true 3D reactive transport simulations should be carried out, eg with models like PHAST (Parkhurst et al. 2004).

## Conclusions

Explicit formulation of microbiological degradation reactions within the PHREEQC framework, offers potentials for simulating intricate biogeochemical processes in nature well. However, there are some real limitations as shown by the two simulation cases presented above. Even in laboratory column studies where more homogeneous and well defined flow conditions to a large extent are approximated, one may have local changes in redox conditions and pore water chemistry in adjacent domains. Effluent sampling may then be composite, and interpreting the geochemistry may then need to involve exchange processes in a dual porosity framework.

Biogeochemical reactions may also be described well in a 3D field situation, but dispersional/diffusional transport effects cannot be simulated sufficiently with the 1D column of PHREEQC alone. To fully benefit from the biogeochemical capabilities of PHREEQC, one needs to couple the model with adequate flow models.

## References


Aagaard P, Tuttle KJ (1996) Protection of groundwater resources against contaminants (The Jens-Olaf Englund Seminar, Gardermoen, 16-18 September)

Aagaard P, Zheng Z, Breedveld G (2001) Computer Modeling of Natural Attenuation of Organic Contaminants and Concomitant Geochemical Processes in Sandy Aquifer Materials. Natural Attenuation of Environmental Contaminants, v. 6(2): 181-188

Alfnes E, Kinzelbach W, Aagaard P (2003a) Numerical simulations of a tracer experiment in an unsaturated soil with dipping layers: Journal of Contaminant Hydrology

Alfnes E, Breedveld G, Kinzelbach W, Aagaard P (2003b) Numerical simulations of coupled contaminant transport and biodegradation in an unsaturated soil with dipping layers: Journal of Contaminant Hydrology

Breedveld GD, Olstad G, Aagaard P (1997) Treatment of Jet Fuel-Contaminated Runoff Water by Subsurface Infitration. Bioremediation Journal, v.1, p.77-88.

Breedveld GD, Sparrevik M, Ådnanes J, Aagaard P (1999) Natural attenuation of jet-fuel contaminated run-off water in the unsaturated zone. In: Alleman and Leeson (eds.), Natural attenuation of chlorinated solvents, petroleum hydrocarbons, and other organic compounds. Battelle Press. pp. 213-218.

Essaid HI, Bekins BA, Godsy EM, Warren E (1995) Simulation of of aerobic and anaerobic biodegradation processes at a crude oil spill site. Water Resources Research, v. 31: 3309-3327.

French HK, van der Zee SEATM & Leijnse A (2001) Transport and degradation of propyleneglycol and potassium acetate in the unsaturated zone. J. Contam. Hydrol. 49: 23-48.

Kindred JS, Celia MA (1989) Contaminant transport and biodegradation: Conceptual model and test simulation. Water Resour. Res. 25: 1149-1160.

Kitterød NO, Finsterle S (2000). Inverse Modeling of Unsaturated Flow Combined with Stochastic Simulation Using Empirical Orthogonal Functions (EOF). In: Groundwater 2000, Eds: Bjerg PL, Engesgaard P, Krom ThD, 2000 A.A. Balkema, Rotteradam, Netherlands, ISBN 90 5809 133 3, pp 103-104

Klonowski MR (2005) Spatial and temporal changes of jet-fuel contamination in an unconfined sandy aquifer. Manuscript.

Klonowski M, Aagaard P, Breedveld GD, Rudolph-Lund K (2002): Natural attenuation of a hydrocarbons plume in an unconfined sandy aquifer, Gardermoen icecontact delta, southeast Norway. Proceedings 11. Seminar on hydrogeology and environmental geochemistry, 7-8.02.2002, Norwegian Geological Survey, Trondheim. pp. 54-55

Knudsen JBS (2003) Reactive transport of dissolved aromatic compounds under oxygen limiting conditions in sandy aquifer sediments. PhD Dissertation, Dept. of Geology, Univ. of Oslo

Lovley DR, Woodward JC, Chapelle FH (1994) Stimulated anoxic biodegradation of aromatic hydrocarbons using Fe(III) ligands. Nature 370: 128-131.

Parkhurst DL, Appelo CAJ (1999) User's guide to PHREEQC (Version 2)--a computer program for speciation batch-reaction, one-dimensional transport, and inverse geochemical calculations. U.S. Geological Survey Water-Resources Investigations Report 99-4259.

Parkhurst DL, Kipp KL, Engesgaard P, Charlton SR (2004) PHAST--A program for simulating ground-water flow, solute transport, and multicomponent geochemical reactions: U.S. Geological Survey Techniques and Methods 6-A8, 154 p

Plummer LN, Wigley TML, Parkhurst DL (1978) The kinetics of calcite dissolution in CO2 water systems at 50 to 600 and 0.0 atm to 1 atm CO2. Am. J. Sci. 278: 179-216.

Rudolph-Lund K, Sparrevik M (1999) Remedial action plan for natural attenuation of contamination at the fire training field at Gardermoen (in Norwegian). NGI report nr. 984138-2. Norwegian Geotechnical Institute, Oslo.

Schäfer D, Schäfer W, Kinzelbach W (1998) Simulations of reactive processes related to biodegradation in aquifers 2. Model application to a column study on organic carbon degradation. J. Contam. Hydrol. 31: 187-209.

Søvik AK, Alfnes E, Breedveld G, French HK, Pedersen TS, Aagaard P (2002) Transport and degradation of toluene and o-xylene in an unsaturated soil with dipping sedimentary layers: Journal of Environmental Quality, v. 31: 1809-1823

Thullner M, Schäfer W (1999) Modeling of a Field Experiment on Bioremediation of Chlorobenzenes in Groundwater. Bioremediation Journal. 3 (3): 247-267.

Tuttle KJ (1997) Sedimentological and hydrogeological characterisation of a raised ice contact delta – complex at Gardermoen Southestern Norway. PhD Dissertation, Dept. of Geology, Univ. of Oslo.

Van Genuchten MTh (1985) A general approach for modeling solute transport in structured soils: IAH Memoirs, v. 17, p. 513-526.

Williamson MA, Rimstidt JD (1994) The kinetics and electrochemical rate determining step of aqueous pyrite oxidation. Geochim. et Cosmochim. Acta. 58: 5443-5454.

Wong WK (2003) Contaminant migration and its risk mapping in the subsurface – the case of Gardermoen, Norway. Ph.D. thesis University of Oslo

Zheng Z, Aagaard P, Breedveld G (2002) Sorption and anaerobic biodegradation of soluble aromatic compounds during groundwater transport. 2. Computer modeling: Environmental Geology, v. 41: 933-941

# Assessing the potential for natural or enhanced in-situ bioremediation at a TCE-contaminated site by coupling process analysis and modeling

F. Aulenta[1], A. Di Fazio[2], M. Leccese[2], M. Majone[1], M. Petrangeli Papini[1], S. Rossetti[4], N. Stracqualursi[3], V. Tandoi[4], P. Viotti[2]

[1] Dept. of Chemistry, University "La Sapienza", P.le Aldo Moro 5, 00185, Rome, Italy
[2] Dept. of Hydraulics, Transportation & Roads, University "La Sapienza", Via Eudossiana, 18, 00184 Rome, Italy
[3] Dept. of Civil Engineering - University "Tor Vergata", Rome Italy, V. di Tor Vergata 110, 00133 Rome, Italy
[4] Water Research Institute (IRSA-CNR), Via Reno 1, 00198, Rome, Italy

## Abstract

Groundwater and soil contamination by chlorinated hydrocarbons has recently become of increasing concern and extensive researches have been conducted to develop technologies for remediating both contaminated groundwaters and soils. In this framework *in-situ* technologies are promising for remediating contaminated groundwater in that they would keep the ecosystem largely undisturbed and would be cost effective.

The aim of the present study is the assessment of natural attenuation potential at a TCE-contaminated site in Northern Italy. This site has a long history of contamination (approximately 50 years) mainly due to industrial activities.

The study illustrates the steps involved in implementing natural or enhanced attenuation screening protocols at this site and represent an outstanding example of effective coupling of process analysis and modeling. In fact microcosm studies properly integrated with modeling results might suggest the feasibility for enhancing *in-situ* reductive dechlorination at the investigated site in order to achieve the stringent legislation limits. Such promising results will be verified through a field test performance before implementing the process at full-scale.

## Keywords

bioremediation; modeling; Chlorinated solvents; Dehalococcoides; Microcosm study; Reductive dechlorination

## Introduction

In-situ enhanced anaerobic reductive dechlorination (RD) is a promising technology for remediation of chloroethane and chloroethene contaminated groundwater. Indeed it is ideally suited for integration into long-term site management programs to address chlorinated solvents dissolved in groundwater. That is because such contaminants often are encountered in the form of DNAPL, which is characterized by a slow release, thus requiring long-term decontamination activities. Consequently active processes (e.g., conventional groundwater pumping and treating) often result costly and quickly reach a point of diminishing returns.

In situ enhanced RD can be accomplished by either enhancing halo respiratory activity of native microbial dechlorinating population (e.g. through the addition of electron donors and/or nutrients to produce favorable reducing conditions), or by inoculating the aquifer with microorganisms that are capable of degrading the target pollutants.

This work is aimed at the implementation of a technical protocol for assessing the suitability of sites for bioremediation of chlorinated solvents. The protocol will be applied at a contaminated site located in Northern Italy where contamination of aquifers by TeCA, PCE, and TCE has been documented.

The protocol to be developed will describe a phased-approach for the assessment of the feasibility of bioremediation, either enhanced or as a part of a natural attenuation objective, at a contaminated site. The "Technical protocol for evaluating natural attenuation of chlorinated solvents in ground water" (EPA/600/R-98/128) and the "Treatability test for evaluating the potential applicability of the reductive anaerobic biological in situ treatment technology (RABITT) to remediate chloroethenes" (US Department of Defence) will be the reference reports for the implementation of this new protocol. The phase-approach will include chemical, hydrogeological, and microbiological characterization of the site, monitoring and modelling of the contamination plume, and field tests.

## Site description and characterization

The protocol is currently being applied at a chlorinated solvent contaminated site located in Northern Italy (Figure 1). This site has a long history of contamination (approximately 50 years) mainly due to industrial activities.

**Fig. 1.** Aerial view of the tested site

In the first phase an extensive review of existing site data (hydrogeological information and aquifer parameters), and sampling and analyses of groundwater at existing wells (20 monitoring wells located in different parts of the plume) has been carried out.

The subsurface below the site shows two clayey layers at different depths which allow to define two aquifer up to about 50 m (Figure 2). The contaminant concentration data collected at the site show that majority of contaminant mass is present in the deep permeable zone named first aquifer probably due to the small thickness of the first clayey layer (0.5-3.5 m) and possible hydraulic connection between the two aquifer. The second as well as the deep aquifers are still not involved in the contamination. Generally in each sampling point two wells were installed at different depth in order to investigate both the shallow and the first aquifer. Groundwater collected from the well located about 20 m down-gradient the contamination source area was mainly contaminated by TeCA (13435 μg/l), PCE (1300 μg/l ), TCE (1970 μg/l), along with *cis*-DCE (290 μg/l and VC at 13 μg/l. The presence of TCE reductive dechlorination daughter products along with an hydraulic conductivity > $10^{-3}$ cm/sec indicated that the site has the potential for natural or enhanced bioremediation.

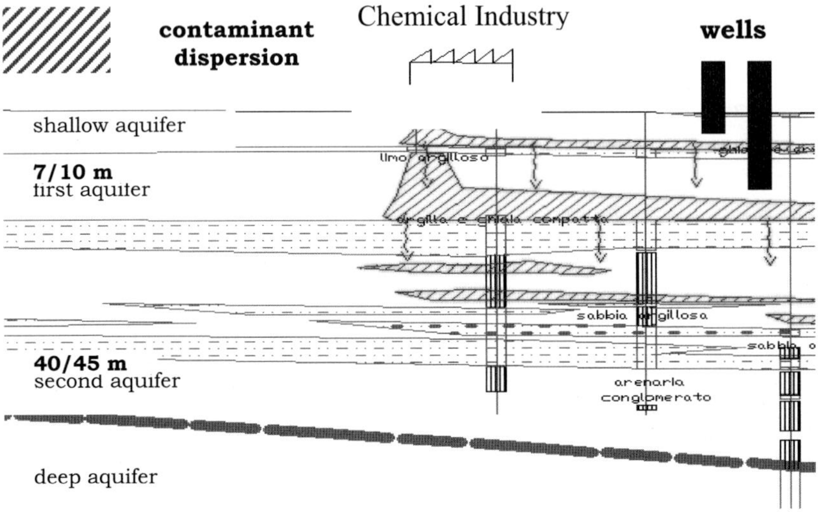


**Fig. 2.** Section of the aquifer system

The second phase of the approach relied on the implementation of a network of monitoring wells (12 piezometers have been installed) close to the source of contamination along groundwater flow direction. This was aimed at increasing the number of groundwater sampling points in order to better monitoring contaminant transport and biodegradation. During this phase aquifer material has been also collected for microcosm studies. The high concentrations detected in the wells all trough the long-term monitoring have suggested the possibility of a DNAPL. Numerical simulations based on different source hypotheses have confirmed that such high levels of contamination could be only explained by the presence of a DNAPL. Future steps will be the implementation of a packer-test monitoring campaign.

## Microcosms studies

The microcosm study was designed to assess the potential for microbial in-situ anaerobic reductive dechlorination (RD) at the contaminated site.
In particular, this study was aimed at evaluating if RD by native populations could be enhanced by the addition of substrates (i.e. yeast extract,

$B_{12}$) and if it could proceed past *cis*-DCE; at evaluating the influence on RD of competitive metabolisms (i.e. nitrate reduction and sulfate reduction); and at evaluating the application of a *Dehalococcoides* spp. containing inoculum for the treatment of the TeCA, PCE and TCE contaminated groundwater.

Microcosms were prepared by using soil and groundwater samples taken from the contaminated site, near the contamination source. Groundwater used for microcosm studies contained TeCA, PCE, TCE at concentration levels previously indicated, nitrate (0.6 mmol/l), and sulfate (5.1 mmol/l) along with trace levels of other chlorinated aliphatic and aromatic compounds (overall COD approx. 30 mg/l). Soil samples were collected by following the procedures described by Fennell et al. (2001), and by Morse et al. (1998). After collection, soil and groundwater were stored in glass jars and maintained in coolers at 4°C until use. For microcosm preparation, the soil samples, groundwater, autoclaved 250-ml serum bottles, gray butyl Teflon-faced stoppers, spatulas, and other material were placed inside an anaerobic glovebox, under nitrogen atmosphere. Thirteen microcosm treatments were set up and triplicate bottles were prepared for each treatment. The experimental conditions for the bottle microcosms are shown in Table 1.

For treatment 1 to 10 (Table 1), 60 g (dry weight) of soil was dispensed in a 250 ml serum bottle, and added with 150 ml of groundwater. Groundwater was previously added with 1 mg/l resazurin (as a redox indicators). After preparation, the bottles were sealed with Teflon-faced butyl rubber stoppers and spiked with the selected electron donor (yeast extract, lactate, butyrate, hydrogen, or none). Each electron donor was added either along with growth factors (i.e. yeast extract at 20 mg/l and vitamin $B_{12}$ at 0.05 mg/l) or without. Yeast extract, lactate, butyrate, and growth factors, were added from a stock solution by using syringes. Hydrogen gas was added in the headspace of the serum bottles, by using gas-tight syringe. After preparation, treatment 1 was autoclaved at 121°C for 1 h (abiotic control). For treatment 11 (bioaugmentation microcosm), 35 ml of a $H_2$-utilizing, PCE-dechlorinating culture containing *Dehalococccoides* spp. (Aulenta et al. 2002; Aulenta et al. 2004) was dispensed in a 250-ml serum bottle and added with 150 ml of groundwater (in the absence of soil). Thereafter, the bottle was sealed and spiked with $H_2$ and growth factors. For treatment 12 and 13, 60 g (dry weight) of soil was dispensed in a 250 ml serum bottle, and diluted with 150 ml reduced anaerobic mineral medium (RAMM). RAMM composition and preparation is that reported by Tandoi et al. (1994). Thereafter the bottle was sealed and spiked with TCE and $H_2$ (treatment 12) or TCE and butyrate (treatment 13). All the microcosms were incubated statically in the dark at room temperature.

**Table 1.** Bottle microcosms experimental conditions and dechlorinating activity observed after 98 days of incubation

| Treat. | Soil and groundwater added with: | Cumulative released chloride[1] (μmol/l) | Dechlorination past cis-DCE |
|---|---|---|---|
| 1 | None (abiotic control) | 0.0 (0.7) | No |
| 2 | None (biotic control) | 12.6 (1.9) | No |
| 3 | None + g.f.[1] | 32.1 (1.6) | No |
| 4 | YE (200 mg/l) + g.f. | 54.2 (3.0) | No |
| 5 | Lactate (3 mmol/l) | 110.0 (11.7) | Yes (mainly VC) |
| 6 | Lactate (3 mmol/l) + g.f. | 157.8 (9.8) | Yes (mainly VC) |
| 7 | Butyrate (3 mmol/l) | 48.7 (20.8) | Yes (mainly VC) |
| 8 | Butyrate (3 mmol/l)+ g.f. | 106.7 (20.2) | Yes (mainly VC) |
| 9 | Hydrogen (3 mmol/l) | 12.9 (0.5) | No |
| 10 | Hydrogen (3 mmol/l) + g.f. | 60.3 (6.2) | No |
| | **Groundwater added with:** | | |
| 11 | Hydrogen (3 mmol/l) + Inoculum[2] + g.f. | 167.8 (14.5) | Yes (VC, ETH) |
| | **Soil added with:** | | |
| 12 | RAMM[3] + TCE + Hydrogen (3mmol/l) + g.f. | 10.6 (0.7) | No |
| 13 | RAMM[3] + TCE + Butyrate (3mmol/l) + g.f. | 48.6 (2.7) | Yes (ETH) |

[1] the chloride cumulative release is the amount of chloride released by the dechlorination processes as calculated (at any time) from the sum of all the measured (by gas-chromatography) dechlorination intermediates, by taking into account their initial and residual chlorination degree. Mean values and standard deviations (in parentheses) of three replicates are shown.
[2] g.f. (growth factors): yeast extract (20 mg/l) and vitamin $B_{12}$ (0.05 mg/l)
[3] the inoculum is a $H_2$-utilizing PCE dechlorinating culture containing Dehalococcoides spp.
[4] RAMM: reduced anaerobic mineral medium

All the electron donors enhanced dechlorination with respect to the biotic control, with butyrate- and lactate- amended microcosms performing the best. For all the tested donors, the addition of trace elements had a beneficial effect on the dechlorinating activity. Evidence for dechlorination past *cis*-DCE was present only in lactate- and butyrate- amended microcosms where VC was the main product along with ethene (ETH) at trace levels.

It is noteworthy that dechlorination was observed also in the presence of nitrate and sulfate. However, added electron donors also promoted the reduction of nitrate and sulfate, with the exception of $H_2$ that did not promote sulfate reduction.

In the microcosm where groundwater was bioaugmented with a PCE-dechlorinating mixed culture containing *Dehalococcoides* spp. chlorinated contaminants (TeCA, PCE, and TCE) were degraded at high rate, and mainly converted to VC, *cis*-DCE and ETH.

In conclusion, results from the microcosm study indicate the potential for enhancing full dechlorination at the chlorinated solvent contaminated site, through a proper addition of a suitable electron donor (e.g. lactate or butyrate) and/or through bioaugmentation with a *Dehalococcoides*-containing culture (Aulenta et al. 2003).

All the microcosms, with the only exception of the abiotic control, showed dechlorinating activity, suggesting the presence of native dechlorinating populations in soil. All the electron donors enhanced dechlorination with respect to the biotic control, with butyrate- and lactate- amended microcosms performing the best. In particular the lactate-amended microcosms showed the shortest lag phase prior to the onset of dechlorination and the highest initial dechlorination rate. For all the tested donors, the addition of growth factors (i.e. yeast extract 20 mg/l and vitamin $B_{12}$) had a beneficial effect on the dechlorinating activity. This suggests that the activity of soil microorganisms was probably limited by the lack of micronutrients. Evidence for dechlorination past *cis*-DCE was present only in lactate- and butyrate- amended microcosms where VC was the main product along with ethene (ETH) at trace levels.

It is noteworthy that dechlorination was observed also in the presence of nitrate and sulfate. However, added electron donors also promoted the reduction of nitrate and sulfate, with the exception of $H_2$ that did not promote sulfate reduction. The finding that in the present study dechlorination occurred in the presence of sulfate and nitrate could be due to that sufficient electron donor was added so minimizing the competition between dechlorination and competing metabolic processes.

All added electron donors promoted the rapid reduction of nitrate, whereas sulfate reduction was promoted only by lactate and butyrate. Neither sulfate nor nitrate reduction was observed in the non-amended biotic control (treatment 2), likely indicating that soil organic carbon was probably not enough to support activity of native microorganisms. Nevertheless, added substrates were effectively utilized by native soil microorganisms. In particular, lactate and butyrate were rapidly converted to acetate, which accumulated up to 6 mM, (and also propionate for lactate), whereas only little acetate was produced from hydrogen ($< 0.4$ mM). Methanogenic activity was not observed in any microcosms probably because of the inhibitory effect of the chlorinated solvents on methanogenic populations. Nevertheless, it is also possible that methanogens were present initially in the soil at such low number (also considering the low soil organic carbon content), hence the development of significant methanogenic activity would have required longer incubation time.

In the microcosm where groundwater was bioaugmented with a PCE-dechlorinating mixed culture containing *Dehalococcoides* spp., chlorinated contaminants (TeCA, PCE, and TCE) were degraded at high rate, and mainly converted to VC, *cis*-DCE and ETH.

In conclusion, results from the microcosm study indicate the potential for enhancing full dechlorination at the chlorinated solvent contaminated site, through a proper addition of a suitable electron donor (e.g. lactate or butyrate) and/or through bioaugmentation with a *Dehalococcoides*-containing culture (Aulenta et al. 2004).

In conclusion, the results of the microcosm study suggest that the potential exists for enhancing full dechlorination at the chlorinated solvent contaminated site, through a proper addition of a suitable electron donor (e.g. lactate or butyrate) and/or through bioaugmentation with a *Dehalococcoides*-containing culture.

## Field test

Final steps of protocol application are the design and execution of a field test for an in situ evaluation of the remediation strategy for the contaminated site.

This implies first selecting a testing location, which meets all technical and administrative screening criteria. Based on the results of the characterization different viable locations have been hypothesized and VISUALMODFLOW simulations have been carried out in order to select the most suitable configuration of the well and monitoring system.

In order to obtain an easier regulatory approval it was decided that the test be located in the hot spot area, which had previously been horizontally confined by a low permeability cement barrier.

The selected area results then completely isolated from the aquifer system; the barrier in fact provides lack of hydraulic connection with the shallow aquifer while the first clayey layer serves as the aquitard.

The proposed field treatability testing system, illustrated in Figure 3, distributes feed solution by forcibly injecting amended groundwater at the head of the testing zone while extracting groundwater near the end of the zone. This technique creates a hydraulic gradient designed to direct the flow of amended groundwater through the test plot.

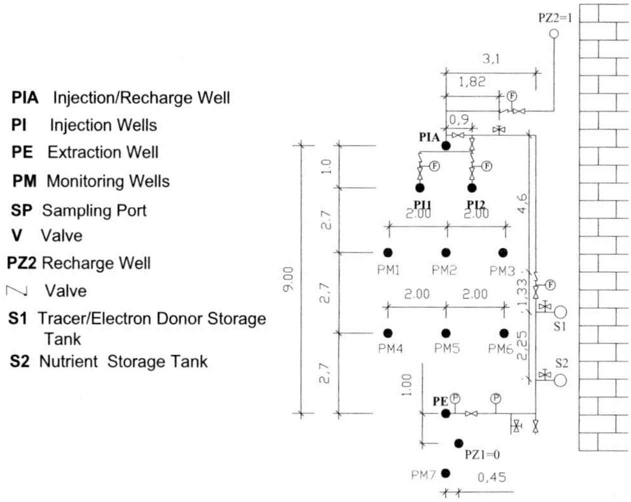


**Fig. 3.** Field test layout (adapted from RABITT)

VISUALMODFLOW simulation (Figure 4) have been carried out in order to determine the size of the area interested by the forced water circulation system. This is necessary in order to calculate the amount of electron donor to be supplied for effectively enhancing the biological activity, thus ensuring a proper process performance.

Based on the site geosystem resulting from investigations evaluation it was decided that the site be represented as a three layered system. Different system configuration have been simulated in order to depict the one more likely able to produce an effective hydraulic control while homogeneously distributing the electron donor.

The behavior of the system has also been investigated, in terms of electron donor dispersion in the system. No consumption from biomass was considered in order to have a conservative estimate of the dispersion area.

Results of the simulations, referred to 150 days continuous injection of lactate in the system, are reported in Figure 5. It is noteworthy that lactate dispersion is confined to a quite limited volume.

The field test project has been completed and realized as shown in Figure 6, after regulatory approval was accomplished.

The results of the simulations are being verified by performing a conservative forced gradient tracer test. The objectives of such test are, in fact, to check the actual hydraulic functioning of the system and to determine the average subsurface tracer residence time.

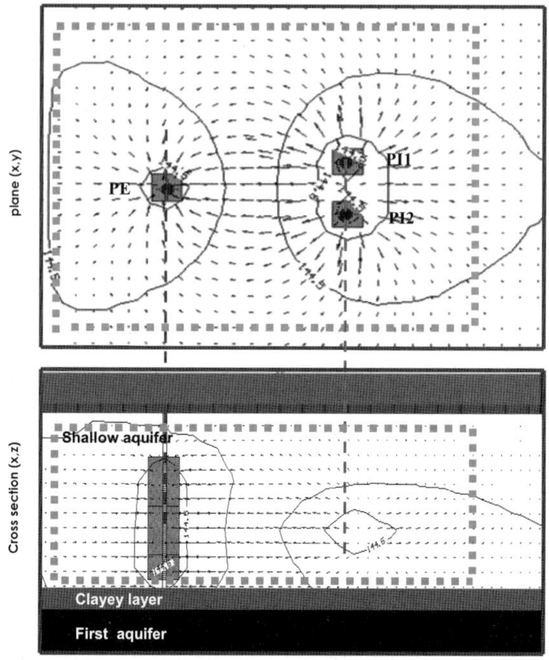


**Fig. 4.** Simulated flow field

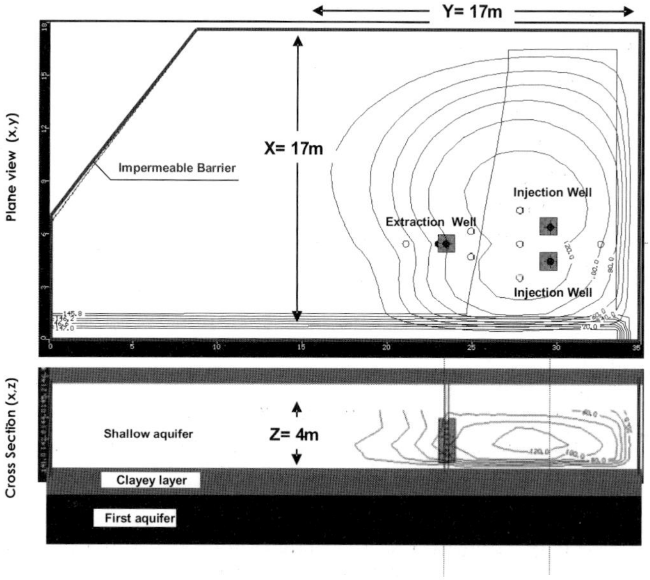


**Fig. 5.** Simulated substrate concentration distribution after 150 days

**Fig. 6.** Overview of the aboveground equipment at sample site.

A concentrated Sodium Chloride solution (200g/l) was used as a tracer and mixed with recirculating water before reinjection in the system. A pre-injection water flood started 24 hours before tracer injection to establish steady-state flow regime in the well field; it was followed by a 3 hours pulse injection. Tests were performed at different recirculation flow rates (13 l/min, 11 l/min, 9 l/min).

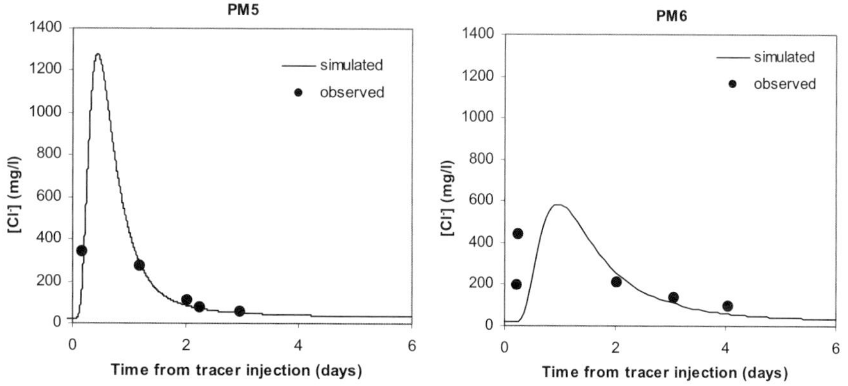


**Fig. 7.** Comparison between simulated and observed BTCs (Q = 9 l/min, $C_0$ = 10.000 mg/l)

Quantifiable concentrations of Chloride appeared at all monitoring locations in a rather short period of time. Observed BTCs for each of the monitoring wells were fitted with VISUALMODFLOW and first results are represented in Figure 7.

Parameters derived from tracer studies proved to be reasonably homogeneous; such information will allow a more realistic simulation of the actual field activity to be performed. In fact protocol application at the sample site will be completed with a field realization aimed at enhancing full dechlorination through a proper addition of a suitable electron donor.

## Conclusions

Assessing the potential for in-situ ENA of chorinated contaminants requires real knowledge of both the processes involved and of the hydrogoelogical behavior of the aquifer. This can only be accomplished by means of a proper coupling of process analysis and modeling with respect to standardized procedures. The development of a protocol has been discussed in this paper and an application has been presented. First results prove that, at the investigated site, ENA has the potential for effectively reducing the level of contamination by chlorinated solvents. However the effectiveness of such a strategy relies on a correct design and management of the treatment system at the field scale. Such aspects are being investigated by performing a field test. The field test is the result of an extensive previous work based on both lab tests and 3D numerical simulations of groundwater flow as well as tracer and substrate transport which have been carried out using standard software such as MODFLOW and MT3D.

## Acknowledgments

This research was financially supported by the "Ministero dell'Ambiente e della Tutela del Territorio" and the National Research Council CNR, GNDRCIE. We gratefully acknowledge also A. Calloni and P. Sala (Provincia di Milano) for providing expert assistance and support for field operations.

## References

Aulenta F, Majone M, Beccari M, Perna L, Tandoi V. (2003a) Enrichment from activated sludges of aerobic mixed-culters capable to degrade vinyl chloride (VC) as the sole carbon source. Annali di Chimica 93, 337-346.

Aulenta F, Giarolli F, Majone M, Tandoi V, Petrangeli Papini M, Beccari M. (2003c) Reductive dechlorination of tetrachloroethene (PCE) to ethene (ETH) in anaerobic sequencing batch biofilm reactor. AIDIC Conference series Vol. 6, pp 25-34 (AIDIC & Reed Business Information) ISBN 0390-2358.

Aulenta F, Bianchi A, Majone M, Petrangeli Papini M, Potalivo M, Tandoi V. (2005) Assessment of natural or enhanced in-situ bioremediation at a chlorinated solvent contaminated aquifer in Italy: a microcosm study. Environment International. 31, 185-190.

Aulenta F, Majone M, Petrangeli Papini M, Beccari M, Tandoi V. (2003d) Enhanced reductive dechlorination of tetrachloroethene to ethene in sediment microcosms. Proceedings of the 2$^{nd}$ International Conference on Remediation of Contaminated Sediments. Battelle. 30 September- 3 October 3003, Venice.

Aulenta F, Majone M, Verbo P, Tandoi V. (2002) Complete dechlorination of tetrachloroethene to ethene in presence of methanogenesis and acetogenesis by an anaerobic sediment microcosm. Biodegradation 13, 411-424.

Aulenta F, Rossetti S, Majone M, Tandoi V. (2004) Detection and quantitative estimation of Dehalococcoides spp. in a dechlorinating bioreactor by a combination of FISH (Fluorescent In Situ Hybridization) and kinetic analysis. Applied Microbiology and Biotechnology, 64, 206-212.

Fennell DE, Carroll AB, Gossett JM, Zinder SH (2001) Assessment of indigenous reductive dechlorinating potential at a TCE-contaminated site using microcosm, polymerase chain reaction analysis, and site data. Environ. Sci. Technol. 35:1830-1839.

ITRC - Interstate Technology and Regulatory Cooperation (1999) Natural Attenuation of Chlorinated Solvents in Groundwater: Principles and Practices (www.itrcweb.org)

Morse JJ, Alleman BC, Gossett JM, Zinder SH, Fennell DE, Sewell GW, Vogel CM (1998) Draft Technical Protocol – A Treatability Test for Evaluating the Potential Applicability of the Reductive Anaerobic Biological In Situ Treatment Technology (RABITT) to Remediate Chloroethenes. US DoD Environmental Security Technology Certification Program. (www.estcp.org.).

Rossetti S, Blackall LL, Majone M, Hughenholtz P, Plumb JJ, Tandoi V (2003) Kinetic and phylogenetic characterization of an anerobic dechlorinating microbial community . Microbiology 149, 459-469

Tandoi V, DiStefano TD, Bowser PA, Gossett JM, Zinder SH (1994) Reductive dehalogenation of chlorinated ethenes and halogenated ethanes by a high-rate anaerobic enrichment culture. Environ. Sci. Technol. 30,973-979.

Wiedmeier et al. (1998) U.S. EPA, Technical protocol for evaluating natural attenuation of chlorinated solvents in ground water, Office of Research and Development, EPA/600/R-98/128.

# Partial source treatment by in-situ technologies – a review of limits, advantages and challenges

S. Grandel, A. Dahmke

Institute of Geosciences, Dept. of Applied Geology,
Christian Albrechts University Kiel, Ludewig-Meyn-Str. 10, 24118 Kiel,
Germany, email: sg@gpi.uni-kiel.de

## Abstract

Removal of contaminant sources or associated residual free phase pools often suffers from a combination of inefficiency, increased risk of contaminant spreading due to uncontrolled mobilization, and/or high treatment costs. The paper gives a brief overview of results from laboratory and field studies where chemical and biochemical in-situ source control actions are evaluated with respect to their efficiency on changes of source emission. The studies focused on the contaminant group of chlorinated ethenes. Chemical approaches aim to mobilize contaminant phases. By now these studies have been mainly executed in the laboratory and only few pilot-scale field studies exist. The results indicate large emission rates at the beginning of the phase displacement but give less information on long-term emission rates. Biostimulation and bioaugmentation approaches revealed increased emission on short time scales and accumulation of cis-Dichloroethylene and Vinylchloride due to incomplete degradation of higher chlorinated solvents in the source zone. On long-term scales emission rates decreased and groundwater plumes were shrinking.

## Keywords

source treatment; in-situ; assessment

## Introduction

Remediation of sites contaminated with organic pollutants often bear the problem of long-term site management or no site-closure due to failures in reaching final cleanup goals. Chlorinated aliphatic hydrocarbons (CAH) are by far the most prevalent organic contaminants in groundwater in Germany (Arneth et al. 1989) and are among the most difficult contaminants to clean up, particularly, when their dense-nonaqueous-phase-liquid (DNAPL) sources remain in the subsurface (Stroo et al. 2003). Advantages of a complete source removal are evident however practical efforts have failed in most cases due to a number of reasons. The localization of the source if at all is possible short time (days to weeks) after the damage occurred, which is hardly the case for most former industrial sites. There, large uncertainties in source delineation exist as even minor subsurface heterogeneities can lead to extremely complex migration pathways and localized entrapment of blobs, ganglia or residual DNAPL (Pankow and Cherry 1996, Poulson and Kueper 1992, Parker et al. 2003). Detection and characterization of DNAPL by point investigations is therefore limited. But even when sources can be located, position and size of the source often hinder complete removal because of financial aspects or technical limitations in consequence partial source treatment is achieved at the best.

In this context innovative in-situ source remediation technologies, which are based on physical, chemical or biological approaches, are discussed as alternative or additional options to conventional measures. However their efficiency (optimization of time, energy, and cost) and effectiveness (ability to reach remediation goals) are hard to assess. Current applications indicate mass removal and/or mass discharge reduction, yet confidential predictions whether and on what time scales they are effective are rare (Stroo et al. 2003). This is reflected by an on-going intense and controversial discussion whether and how to use these in-situ approaches and by which performance goals and metrics they can be assessed in the field (ITRC 2004; U.S.EPA 2003).

This paper aims to give an overview of the current status on performance assessment and application of innovative in-situ source treatment technologies to treat DNAPL in the saturated zone. It represents a literature study, compiling about 100 publications from the U.S. Environmental Protection Agency (EPA), the U.S. Interstate Technology & Regulatory Council (ITRC), and international research journals primarily from the years 2000-2004. After a summary on the current discussion about definition and evaluation of in-situ source treatment benefits, the various technologies are discussed in terms of their applicability, estimated effectiveness, and compatibility with other remediation approaches.

## Source zone characterization and treatment

### Source treatment benefits – expectations versus reality

Sources are described as "contaminated material that acts as a reservoir (either stationary or mobile) for the continued migration of contamination to surrounding environmental media" (U.S.EPA 2003). The designation of source treatment benefits has been subject of significant on-going technical, scientific and policy debates. Benefits include (1) mitigating future potential for human health through long-term reduction of volume, toxicity and mobility of the DNAPL, (2) mitigating future environmental impacts, (3) reducing the duration and cost of subsequent remediation technologies, (4) reducing the life-cycle cost of site cleanup (U.S.EPA 2003, ITRC 2004). These benefits are achieved if the source treatment results in reduction of (1) DNAPL mobility, (2) longevity of groundwater remediation, and/or (3) of the rate of mass discharge from the source zone. Adverse impacts may result in undesirable changes in DNAPL architecture, increase in solubility of dissolved-phase constituents, production of more soluble (toxic) daughter products, or undesirable changes in the physical, geochemical or microbial conditions of the source or adjacent plume zone (ITRC 2004).

Assessment of magnitude and probability of desired or adverse effects during in-situ DNAPL treatment is challenging, due to the inherent uncertainties on DNAPL architecture and inventory. Even small amounts of residual DNAPL can act as long-term sources of the groundwater, indicating that removal of the vast majority of the DNAPL will be necessary to achieve significant near-term improvements in groundwater quality (Sale and McWhorter 2001, Eberhardt and Grathwohl 2002, Grathwohl 2001, Johnson and Pankow 1992). Reduction of source discharge by factors of 2-10 may thus have little impact on the net present value of a site due to a still existing plume for hundreds of years (Stroo et al. 2003), which might be rated as no real benefit with respect to future land use. Apart from active source treatment composition and mass of source discharge can vary naturally, e.g. DNAPL wettability is changed by aging (Dwarakanath et al. 2002, Tuck et al. 1998), or fluctuations in groundwater levels (Aral and Liao 2002, Broholm et al. 1999).

A decrease in source discharge as a direct consequence of source treatment is only achieved by a distinct reduction of the reactive surface of the contaminant phase. This has been observed in lab experiments (e.g. Schwille 1988) but can badly be predicted by means of models particularly in terms of long-term releases and large NAPL volumes (Imhoff et al. 2003a). Current models used for predictions on DNAPL migration and quantitative remediation impacts are often not sufficiently robust or require data that are difficult to obtain, e.g. long-term performance data (Stroo et al. 2003). Hence model assumptions of a constant contaminant discharge from the source often used in model approaches to simulate plume behavior are at the most a simple, not necessarily conservative but in most cases not at all a correct boundary condition. Sensitivity analyses indicate that source-zone architecture is a primary factor governing mass bulk transfer and source longevity (e.g. Sale and McWhorter 2001). Despite the fact that the importance of physical factors or chemical processes on NAPL movement are theoretically known, and physico-chemical constants are available (e.g. Pankow and Cherry 1996, Mackay et al. 1991), simulation of DNAPL migration on a field scale is hampered by missing descriptions of subscale distributions and up-scaling "routines" (Khachikian and Harmon 2000). In this context first results from a model approach using a modified form of inspectional analysis to scale NAPL movement observed in laboratory models to much larger field scales seem to be promising (Imhoff et al. 2003a).

All these mentioned "scientific" uncertainties are regarded as significant barriers in decision making and legal acceptance of in-situ source treatment options (U.S.EPA 2003). Potential metrics for performance assessment and monitoring tools are proposed by the ITRC (2004), they include assessment of treatment progress, but also evaluation of mass reduction and source treatment impact (Table 1). Evaluation of the two latter is the critical feature due to the inherent problems already mentioned and methodical limitations (see Table 1 and following section) however comparable uncertainties exist for performance assessment of conventional approaches.

## Methods for source zone delineation and characterization

Source zone detection and delineation is the major challenge to characterize the volume and extent of DNAPL releases to the subsurface and hence to design and assess in-situ source treatment measures. This is hard enough in unconsolidated materials but extremely complex in bedrock aquifers, which is reflected by the comparable small amount of bedrock sites where source remediation is applied (11%, Navy Survey Study published by Lebron (2004), cited in ITRC (2004)). A key practical observation is that NAPL in the surface environment cannot be effectively located using the "1% of solubility" rule-of-thumb (U.S.EPA 1992) and observed aqueous concentrations (Khachikian and Harmon 2000, Kram et al. 2001). Given the heterogeneity typical of field sites, the probability of catching the DNAPL by few point values is quite small, which is the weakness of expensive conventional investigation methods via soil-gas analyses, core or well sampling (Table 1). In addition, like all intrusive investigation methods, drilling may cause cross-contamination or establish new migration-pathways. As source investigation has always been an important issue, new investigation technologies are continually being developed and tested (ITRC 2003a, Kram et al. 2001, 2002, U.S.EPA 2004b, Parker et al. 2003).

Geophysical techniques i.e. ground penetrating radar, cross-well radar, electrical resistance tomography, vertical induction profiling, or high resolution seismic 3D reflection (Kram et al. 2001; Temples et al. 2001) provide information on changes in the matrix and geologic structure, but not on the matrix itself and are thus not able to detect DNAPL directly. Correlation to geologic information by means of e.g. borehole data enables a spatial delineation on stratigraphics or preferential flow paths (Kausar 2003). Typical spatial resolution is in the order of meters to tens of meters (Kram et al. 2001), which might preclude determination of ganglia location for most cases.

The various direct-push techniques like e.g. Geoprobe systems, Precision Sampling Multichannel Sampler, cone penetrometer etc. (Kram et al. 2001) opened up the chance for a more cost-effective, faster and numerous sampling of groundwater and soil compared to conventional drilling techniques. These techniques enable discrete sampling or supply in-situ readings of groundwater or soil quality by combining the direct push-apparatus with probe or sensor techniques. This allows for direct using e.g. video imaging, or Raman spectroscopy or indirect using e.g. permeable membrane sensors, hydro sparge, or fluorescence techniques evidence of DNAPL (ITRC 2003a, U.S.EPA 2004b). The approaches are limited by subsurface features (bedrocks, boulders) and comprise the uncertainties connected to point measurements regarding spatial integration.

**Table 1.** Potential categories of performance metrics and methods to assess DNAPL zone treatment (according to ITRC (2004)).

**Source treatment progress**

- *decrease in soil concentration* (drilling, direct-push)
  => need for large numbers of samples, measurement errors, potential errors in interpolation, changes may not be due to treatment
- *decrease in groundwater concentration* (wells, multi-level-sampling, passive diffuse sampling)
  => potential for high variability, measurement errors, interpolation errors, changes may not be due to treatment
- *decrease in soil vapor concentration* (soil vapor extraction)
  => changes may not be due to treatment

**Source mass reduction**

- *mass extracted* (ex-situ measurement of waste streams)
- *mass destroyed in-situ* (indicators of breakdown products: Cl-concentration., C-14 isotopes)
  => background Cl-conc. often too high to detect differences, isotope signals often too weak, data set too small for mass balances
- *mass remaining* (soil cores, PITT)
  => uncertainties connected to source detection and delineation

**Source treatment impact**

- *decrease in toxicity* (contaminant analyses)
  => prediction of long-term effects, availability and interpretation of tox-tests
- *decrease in mobility* (determine NAPL saturation e.g. via PITT)
  => limitations in measuring saturation or source discharge in the field
- *decrease in plume loading* (mass flux via transects or integrated pump test)
  => so far few field applications, sophisticated technologies

Partitioning tracer tests (PITT) are discussed as an option to determine mass and volume of a DNAPL (Table 1). Alcohols or sulfur hexafluorid ($SF_6$) have been used to determine DNAPL residual saturations in laboratory tests (Istok et al. 2002, Wilson and Mackay 1995). Another potential tracer is the naturally occurring Radon-222, which is retarded by organic phases and is sensitive to changes in DNAPL saturation in space and time (Davis et al. 2002, 2003, Kram et al. 2001). However, quantitative estimates of DNAPL volume in the field may be subject to large errors, due to uncertainties regarding tracer interaction effects (Imhoff et al. 2003b), sorption or degradation of tracers (Brooks et al. 2002), and differing subsurface behavior of tracers and contaminants (Brooks et al. 2004). In addition uncertainties in prediction of residual saturation arise from misleading assumptions of homogeneous NAPL distribution, or linear equilibrium partitioning of tracers (Istok et al. 2002). This is also to be considered when PITT are used for assessment of remediation performance (Brooks et al. 2004, Cain et al. 2000, Dugan et al. 2003, Meinardus et al. 2002, Noordman et al. 2000). Currently field experiences on tracer tests are rather restricted and questions on test design under field conditions warrant further investigation (Field 2003).

In summary there are numerous techniques for source investigation (Kram et al. 2001), which should be used complementary to raise the probability of a successful source location and delineation. The advantage of the combined direct-push & probe techniques lies in their speed of advance and sensitivity to the type of soil matrix but bear the disadvantage of point information. Geophysical methods give an integrated picture of the subsurface, but do measure changes in matrix properties rather than chemical makeup (U.S.EPA 2004b), so confirmation by data from intrusive methods is required. Cost comparisons based on model simulations indicate that innovative approaches are generally more efficient than conventional baseline approaches, with direct-push techniques being less expensive than PITT or geophysical techniques (Kram et al. 2002).

## Innovative in-situ source treatment technologies

Innovative in-situ source treatment techniques include thermal treatment, chemical extraction, chemical oxidation, chemical reduction, or enhanced bioremediation. In-situ source treatment has been primarily applied in the United States, with 364 projects reported till 2002 (U.S.EPA 2004c), thereby established technologies i.e. soil vapor extraction were used at 60% and innovative techniques at about 30% of the sites (thermal treatment <7%, bioremediation 13%, flushing 4%, and chemical treatment 7%).

In-situ thermal treatment technologies are used to increase the mobility of DNAPL. The major types are steam injection (SEE), electrical resistive heating (ERH), and thermal conductive heating. They differ in terms of the process used to deliver thermal energy to the subsurface (Table 2) and are based on the fact that VOC under pressure and immersed in water have boiling points below the one of water (ITRC 2004). Vapors formed by heating are generally captured by a soil vapor extraction system, which should be designed to prevent lateral displacement of contaminated fluids outside of the treatment zone. Application is limited by the ability to deliver the heat source below ground surface, reported applications range from 3-48m (U.S.EPA 2004a). Case studies have been reported for about 40 sites, more than 50% were full scale, 50% ERH types, and energy costs accounted typically for <30% of the total project costs (U.S.EPA 2004a).

**Table 2.** Types and principles of innovative in-situ source treatment technologies. (References: [1)] U.S.EPA (2004a), [2)] ITRC (2003b), [3)] Crimi and Siegrist (2003), [4)] Seol et al. (2003), [5)] AFCEE and NFESC (2004)).

**Thermally enhanced remediation** [1)]

- *Steam enhanced extraction*: direct heating by injection of steam or hot air, most effective in zones of moderate to high permeability ($K_f$ $10^{-5}$ m/s)
- *Electrical resistance heating*: heating results from natural resistance of a geologic medium to passage of an electrical current ($K_f$ $10^{-5}$ to $10^{-10}$ m/s)
- *Thermal conduction heating*: heating by conduction from electrically powered thermal wells, mainly applied in soils, up to 500°C can be reached

**Chemical extraction**

- *surfactant flushing*: anionic or nonionic chemical compounds, (1) solubilizing surfactant flooding (i.e. increase of apparent aqueous solubility of the contaminant), (2) mobilizing surfactant flooding (lowering of DNAPL-water interfacial tension) [2)]
- *cosolvent flushing*: generally alcohols, alter properties of solution interfaces and are often combined with surfactants, while surfactants result in micellar solubilization (emulsions), addition of cosolvents produces "true" solutions [2)]

**Chemical oxidation**

- *Permanganate*: commercially available, used as $KMnO_4$ or Na $MnO_4$, might include metal impurities [3, 4)]
- *Fenton's reagent*: mixture of hydrogen peroxide, a catalyst (ferrous sulphate) and an optimum pH adjuster (sulphuric acid) [4)]

**Substrates used for enhanced bioremediation** [5)]

- *Soluble*: lactate, butyrate, methanol, ethanol, sodium benzoate, molasses, high fructose corn syrup
- *Viscous fluids*: HRC® (Hydrogen Release Compounds), vegetable oils
- *Low-viscosity fluids*: vegetable oil emulsions
- *Experimental*: whey (soluble), chitin (solid), hydrogen (gas), humic acids (electron shuttle)

Chemical extraction approaches include surfactant-enhanced aquifer remediation (SEAR) or cosolvent (alcohol) flushing, which differ in the mechanisms of extraction provided by the solutions (Table 2), but result both in mobilization and/or solubilization of the DNAPL (ITRC 2003b). A selected alcohol is not as effective as a carefully selected surfactant at dissolving DNAPL, an alcohol flood thus requires many more pore volumes of flushing solution (ITRC 2003b). Solubilization can be increased by linker-molecules which augment interactions between surfactant and oil or water phase (Acosta et al. 2002), or combined cosolvent/air-injection to enhance contact of cosolvent and DNAPL (Jeong et al. 2002). Downward migration of mobilized DNAPL should be prevented by extraction design, artificial walls or natural barriers. Another option deals with density modification (Kostarelos et al. 1998, Shook et al. 1998), where buoyancy forces of the DNAPL are modified by flushing the subsurface with emulsions (Ramsburg and Pennell 2002, Ramsburg et al. 2003, Yan et al. 2003) or dense brine solutions (Hill III et al. 2001, Miller et al. 2000). Chemical extraction techniques are mainly applied in laboratory (e.g. Jayanti et al. 2002, Taylor et al. 2001, 2004) and pilot-scale studies. The latter showed enhanced solubilization, e.g. aqueous PCE-concentrations were up to 22 times higher compared to pure water flushing, and removal effectiveness ranged from 48-64% (Jawitz et al. 2000, Brooks et al. 2004, Tick et al. 2003). Field application require detailed preliminary information e.g. capillary desaturation tests for selection of appropriate surfactant (Sabatini et al. 2000), or information on background concentrations of anions or cations which might significantly alter the behavior of the surfactant formulation (Field et al. 2000, ITRC 2003b). Primary factors that have limited field demonstrations from proceeding to full-scale applications have been concerns regarding costs of disposal of the effluent and regulatory permitting concerns (ITRC 2003b).

Reduction of source mass can be achieved by in-situ chemical oxidation, chemical reduction or enhanced anaerobic bioremediation. The potential benefit of an in-situ chemical oxidation (ISCO) approach is the rapid and extensive reaction of the oxidant i.e. Fenton's reagent, potassium permanganate, or ozone (Table 2) with various organic contaminants of concern (U.S.EPA 1998). This in turn might result in a need for large quantities of oxidant chemicals due to an unproductive consumption with other available aquifer reductants. ISCO is furthermore limited by process-induced geochemical effects like altered pH-, Eh-values or changes in aquifer permeability caused by Mn oxide precipitation. The latter was observed in laboratory and small field cell tests by Lee et al. (2003), Li and Schwartz (2004b) in contrast to Crimi and Siegrist (2003). Results from Nelson et al. (2001) suggest that decrease in porosity and permeability is unlikely in a neutral or near-neutral carbonate aquifer. Mn oxides could be dissolved by a combined oxidant/organic acid (e.g. EDTA, oxalate and citrate acid) treatment, tested in a tank experiment by Li and Schwartz (2004a). They observed a 100% contaminant removal after 4 flushings. Other ISCO lab studies indicate CAH mass removal of 40-60% (Lee et al. 2003, Schnarr et al. 1998), which was determined by mass balance calculations. This seems not applicable in the same line for field applications, where the measurement of $Cl^-$ concentrations as reaction tracer (MacKinnon and Thomson 2002, Schnarr et al. 1998), or isotope fractionation factors are suggested (Hunkeler et al. 2003, Poulson and Naraoka 2002). Currently this technique is restricted to few small field applications.

Applications suffer mainly from efficiency problems (Seol et al. 2003), i.e. oxidants are consumed during delivery, or reaction products tend to plug the porous medium, suggesting that ISCO performs better in removing residual than pooled DNAPL (e.g. Li and Schwartz 2004a).

Nanoscale zero-valent iron (NZVI), bimetallic (Pb/Fe) nanoscale particles (Lien and Zhang 2001, Zhang 2003), emulsified nanoscale iron particles (EZVI), i.e. nanoscale $Fe^0$ in an aqueous media, surrounded by a hydrophobic oil-membrane (Cho et al. 2002, Geiger et al. 2003), or granular iron (Wadley et al., accepted) may be used for chemical reduction of DNAPL. Although promising these approaches have so far mainly been tested in laboratory studies. No or few information is available on their long-term stability, clogging or passivation effects during field application. Similar to the other injection techniques, design and evaluation of $Fe^0$ colloid distribution under field conditions is challenging. Direct subsurface injection can be achieved under gravity-feed or pressurized conditions, e.g. pressure pulsing techniques, pneumatic fracturing, hydraulic fracturing or direct-push injection (Elliott and Zhang 2001, Quinn et al. 2004). Direct push-injection seems feasible for small sites, whereas pressure techniques might deform EZVI (Quinn et al. 2004). A first field demonstration indicates that TCE was reduced where EZVI was present, but DNAPL mobilization could not be excluded (O'Hara et al. 2004).

Source mass removal might be also achieved by accelerating DNAPL dissolution and then stimulating biodegradation of the dissolved contaminants, which is recorded under "ENA - Enhanced Natural Attenuation" (AFCEE and NFESC 2004, DiStefano et al. 2001, McNab Jr. et al. 2000, Vera et al. 2001, Yang and McCarty 2002). ENA includes injection of electron donors in terms of fermentable substrates or hydrogen release compounds (Table 2), furthermore nutrients or dechlorinating bacteria (bioaugmentation) can be added to the subsurface. The latter has been discussed as an option both for groundwater and source remediation, since CAH degrading cultures showed no or few inhibitory effects under saturated PCE or TCE concentrations (Adamson et al. 2003, Yang and McCarty 2000). Anaerobic dechlorination takes place in the aqueous phase and does not directly attack DNAPL mass however injected organic substrates may lower the interfacial tension between DNAPL and groundwater (AFCEE and NFESC 2004). Laboratory studies indicate that ENA causes enhanced PCE removal by up to a factor of 16 compared to effects of dissolution alone in abiotic reactors (Carr et al. 2000, Cope and Hughes 2001). Dechlorination beyond cis-DCE may be limited to regions down gradient of the source zone as degradation of PCE is thermodynamically and kinetically more favorable (Adamson et al. 2003, Maymo-Gatell et al. 2001, Zheng et al. 2001). Despite the technical basis and reported occurrence of ENA at a limited number of sites (Payne et al. 2001, Sorenson 2003), the real significance of this technique and the potential to engineer it are not fully understood and may be limited (for more details see: AFCEE and NFESC (2004)).

Apart from uncertainties with respect to the effectiveness of in-situ source treatment techniques, their compatibility with subsequent passive plume remediation actions, like monitored natural attenuation (MNA) or reactive barriers is hardly known or investigated although source treatment in general is a legally recommended precondition to apply MNA (U.S.EPA 1999). Looney and Vangelas (2004) or Wiedemeier and Haas (2002) suggest a benefit or enhancement of the designated NA processes in conjunction with in-situ source treatment, except for a negative impact of chemical oxidation on the degradation of highly chlorinated CAH. According to calculations by Christ et al. (in press) source longevity may be reduced by as much as an order-of-magnitude when physical-chemical source-zone treatment is coupled with reductive dechlorination. These assumptions are supported by few field investigations. TCE degradation was inhibited by the lowered pH-value due to treatment with Fenton's reagent (Kastner et al. 2000). O'Hara et al. (2004) report of stimulated TCE degradation after EZVI-injection where the oil served as electron donor, yet abiotic and biological proportions in degradation could not be distinguished. Ethanol which remained to some extent in the subsurface after a cosolvent flushing field test also stimulated biodegradation processes (Jawitz et al. 2000, Mravik et al. 2003, Tick et al. 2003). Examinations of microbial activity after 6 weeks of steam enhanced extraction showed high levels of microbial activity following cooling, including mesophilic populations capable of biodegradation (Richardson et al. 2002).

## Summary and Conclusions

Rising costs and uncertainties associated with long-term management of DNAPL sites have resulted in a reassessment to whether or not source remediation is efficient and possible. Due to the innovation of more rapid and cost-effective field investigation techniques and their suitable combinations, detection and delineation of source location and architecture and thus in-situ source remediation became more feasible. Most associated processes are theoretically understood, yet transferability to field conditions and reliable model prognoses require more knowledge about the relative importance of rate-limiting mechanisms compared to scaling effects and aquifer heterogeneities. All in-situ approaches indicate an increase in source discharge on short time scales. Long term prognoses for the chemical approaches are by now not available. First results from biochemical studies indicate a decrease in source discharge on longer time scales, which would result in the reduction of groundwater plume longevity. This in return is a relevant regulatory criterion for the acceptance of natural attenuation concepts. However the effects of in-situ source treatment on such subsequent measures have to be better understood.

Knowledge on the strengths and weaknesses of the various in-situ treatment approaches and corresponding performance assessment is growing along with the increasing number of field experiences. Remaining questions exist in terms of assessment of long-term effects of source treatment, tools to measure mass fluxes and mass release rates, and tools to assess and predict performance. Research to address these questions is in progress at least for sites located in unconsolidated aquifers (U.S.EPA 2004d), whereas deep and/or fractured aquifers are to a large extent underrepresented.

## References

Acosta E, Tran S, Uchiyama H, Sabatini DA, Harwell JH (2002) Formulating chlorinated hydrocarbon microemulsions using linker molecules. ES&T 36:4618-4624

Adamson DT, McDade JM, Hughes JB (2003) Inoculation of a DNAPL source zone to initiate reductive dechlorination of PCE. ES&T 37:2525-2533

AFCEE, NFESC (2004) Principles and Practices of Enhanced Anaerobic Bioremediation of chlorinated solvents. Report Nr. 022/738863/28.doc.

Aral MM, Liao B (2002) Effect of groundwater table fluctuations on LNAPL thickness in monitoring wells. Env. Geology 42:151-161

Arneth J-D, Milde G, Kerndorff H, Schleyer R (1989) The 15 most frequently detected organic compounds in groundwater at waste disposal sites in Germany and the U.S. In: Baccini P (ed) The Landfill. Springer-Verlag, New York

Broholm K, Feenstra S, Cherry JA (1999) Solvent release into a sandy aquifer. 1. Overview of source distribution and dissolution behaviour. ES&T 33:681-690

Brooks MC, Annable MD, Rao PSC, Hatfield K, Jawitz JW, Wise WR, Wood AL, Enfield CG (2002) Controlled release, blind tests of DNAPL characterization using partitioning tracers. JCH 59:187-210

Brooks MC, Annable MD, Rao PSC, Hatfield K, Jawitz JW, Wise WR, Wood AL, Enfield CG (2004) Controlled release, blind test of DNAPL remediation by ethanol flushing. JCH 69:281-297

Cain RB, Johnson GR, McCray JE, Blanford WJ, Brusseau M (2000) Partitioning tracer tests for evaluating remediation performance. Ground Water 38:752-761

Carr CS, Garg S, Hughes JB (2000) Effect of dechlorinating bacteria on the longevity and composition of PCE-containing nonaqueous phase liquids and equilibrium dissolution conditions. ES&T 34:1088-1094
Cho Y-L, McClements DJ, Decker EA (2002) Ability of surfactant micelles to alter the physical location and reactivity of iron in oil-in-water emulsions. J Agric Food Chem 50:5704-5710
Christ JA, Ramsburg, CA, Abriola LM, Pennell KD, Löffler FE (in press) Coupling aggressive mass removal with microbial reductive dechlorination for remediation of DNAPL source zones - a review and assessment. Environmental Health Perspectives
Cope N, Hughes JB (2001) Biologically-enhanced removal of PCE from NAPL source zones. ES&T 35:2014-2021
Crimi ML, Siegrist RL (2003) Geochemical effects on metals following permanganate oxidation of DNAPLs. Ground Water 41:458-469
Davis BM, Istok JD, Semprini L (2002) Push-pull partitioning tracer tests using radon-222 to quantify non-aqueous liquid contamination. JCH 58:129-146
Davis BM, Istok JD, Semprini L (2003) Static and push-pull methods using Radon-222 to characterize dense nonaqueous phase liquid saturations. Ground Water 41:470-481
DiStefano TD, Barai R, Duran M, Speece RE (2001) A comparison of complex electron donors for anaerobic dechlorination of PCE. Biorem. J. 5:131-143
Dugan PJ, McCray JE, Thyne GD (2003) Influence of a solubilty-enhancing agent (cyclodextrin) on NAPL-water partition coefficients, with implication for partitioning tracer tests. Water Resour. Res. 39:1123
Dwarakanath V, Jackson RE, Pope GA (2002) Influence of wettability on the recovery of NAPLs from alluvium. ES&T 36:227-231
Eberhardt C, Grathwohl P (2002) Time scales of organic contaminant dissolution from complex source zones: coal tar pools vs. blobs. JCH 58:45-66
Elliott DW, Zhang W-X (2001) Field assessment of nanoscale bimetallic particles for groundwater treatment. ES&T 35:4922-4926
Field JA, Sawyer TE, Schroth MH, Humphrey MD, Istok JD (2000) Effect of cation exchange on surfactant-enhanced solubilization of trichloroethene. JCH 46:131-149
Field MS (2003) A review of some tracer-test design equations for tracer-mass estimation and sample collection frequency. Env. Geology 43:867-881
Geiger CL, Clausen CA, Brooks K, Clausen C, Huntley C, Filipek L, Reinhart DR, Quinn J, Krug T, O'Hara S, Major D (2003) Nanoscale and microscale iron emulsions for treating DNAPL. In: Henry SM, Warner SD (ed) Chlorinated solvent and DNAPL remediation: Innovative strategies for subsurface cleanup. ACS Symposium Series 837:132.140
Grathwohl P (2001) Zeitskalen der Schadstofflösung und -desorption: Natural Attenuation im Abstrom persistenter Schadstoffquellen. F. Geoök. 12:28-34
Hill III EH, Moutier M, Alfaro J, Miller CT (2001) Remediation of DNAPL pools using dense brine barrier strategies. ES&T 35:3031-3039

Hunkeler D, Aravena R, Parker BL, Cherry JA, Diao X (2003) Monitoring oxidation of chlorinated ethenes by permanganate in groundwater using stable isotopes: laboratory and field studies. ES&T 37:798-804

Imhoff PT, Mann AS, Mercer M, Fitzpatrick M (2003a) Scaling DNAPL migration from the laboratory to the field. JCH 64:73-92

Imhoff PT, Pirestani K, Jafarpour Y, Spivey KM (2003b) Tracer interaction effects during partitioning tracer tests for NAPL detection. ES&T 37:1441-1447

Istok JD, Field JA, Schroth MH, Davis BM, Dwarakanath V (2002) Single-well "Push-Pull" partitioning tracer test for NAPL detection in the subsurface. ES&T 36:2708-2716

ITRC (2003a) An introduction to characterizing sites contaminated with DNAPLs

ITRC (2003b) Technical and regulatory guidance for surfactant/cosolvent flushing of DNAPL source zones

ITRC (2004) Strategies for monitoring the performance of DNAPL source zone remediation

Jawitz JW, Sillan RK, Annable MD, Rao PSC, Warner K (2000) In-situ alcohol flushing of a DNAPL source-zone at a dry cleaner site. ES&T 34:3722-3729

Jayanti S, Britton LN, Dwarakanath V, Pope GA (2002) Laboratory evaluation of custom-designed surfactants to remediate NAPL source zones. ES&T 36:5491-5497

Jeong S-W, Wood AL, Lee TR (2002) Enhanced contact of cosolvent and DNAPL in porous media by concurrent injection of cosolvent and air. ES&T 36:5238-5244

Johnson RL, Pankow JF (1992) Dissolution of dense chlorinated solvents into groundwater. 2. Source functions for pools of solvent. ES&T 26:896-901

Kastner JR, Domingo JS, Denham M, Molina M, Brigmon R (2000) Effect of chemical oxidation of subsurface microbiology and Trichloroethene (TCE) biodegradation. Biorem. J. 4:219-236

Kausar T (2003) 3D stochastic modelling of aquifer properties in a fluvial setting based on geophysical measurements in the framework of the RETZINA project. PhD, Eberhardt-Karls- Universität Tübingen

Khachikian C, Harmon TC (2000) Nonaqueous phase liquid dissolution in porous media: current state of knowledge. Transport in Porous Media 38:3-28

Kostarelos K, Pope GA, Rouse BA, Shook GM (1998) A new concept: the use of neutrally-bouyant microemulsions for DNAPL remediation. JCH 34:383-397

Kram ML, Keller AA, Rossabi J, Everett LG (2001) DNAPL characterization methods and approaches, Part 1: Performance comparisons. GWMR 21:109-123

Kram ML, Keller AA, Rossabi J, Everett LG (2002) DNAPL characterization methods and approaches, Part 2: Cost comparison. GWMR 22:46-61

Lee ES, Seol Y, Fang YC, Schwartz FW (2003) Destruction Efficiencies and Dynamic of Reaction Fronts Associated with the Permanganate Oxidation of Trichlorethylene. ES&T 37:2540-2546

Li XD, Schwartz FW (2004a) DNAPL remediation with in situ chemical oxidation using potassium permanganate. II. Increasing removal efficiency by dissolving MN oxide precipitates. JCH 68:269-287

Li XD, Schwartz FW (2004b) DNAPL remediation with in-situ chemical oxidation using potassium permanganate. Part I. Mineralogy of Mn oxide and its dissolution in organic acids. JCH 68:39-53

Lien HL, Zhang WX (2001) Nanoscale iron particles for complete reduction of chlorinated ethenes. Colloids and Surfaces A: Physicochemical and Engineering Aspects 191:97-105

Looney B, Vangelas KM (2004) Compatibility of alternative chlorinated solvent source treatment strategies with monitored natural attenuation, WSRC-MS-2004-00236:1-21

Mackay DM, Shiu WY, Maijanen A, Feenstra S (1991) Dissolution of non-aqueous phase liquids in ground water. JCH 8:23-43

MacKinnon LK, Thomson NR (2002) Laboratory-scale in situ chemical oxidation of a perchloroethylene pool using permanganate. JCH 56:49-74

Maymo-Gatell X, Nijenhuis I, Zinder S (2001) Reductive dechlorination of cis-1,2-Dichloroethene and Vinyl Chloride by "Dehalococcoides ethenogenes". ES&T 35:516-521

McNab Jr. WW, Ruiz R, Reinhard M (2000) In-situ destruction of chlorinated hydrocarbons in groundwater using catalytic reductive dehalogenation in a reactive well: testing and operational experiences. ES&T 34:149-153

Meinardus HW, Dwarakanath V, Ewing J, Hirasaki GJ, Jackson RE, Jin M, Ginn JS, Londergan JT, Miller CA, Pope GA (2002) Performance assessment of NAPL remediation in hetereogeneous alluvium. JCH 54:173-193

Miller CT, Hill III EH, Moutier M (2000) Remediation of DNAPL-contaminated subsurface systems using density-motivated solubilization. ES&T 34:719-724

Mravik SC, Sillan RK, Wood AL, Sewell GW (2003) Field evaluation of the solvent extraction residual biotreatment technology. ES&T 37:5040-5049

Nelson MD, Parker BL, Al TA, Cherry JA, Loomer D (2001) Geochemical reactions resulting from in situ oxidation of PCE-DNAPL by $KMnO_4$ in a sandy aquifer. ES&T 35:1266-1275

Noordman WH, De Boer GJ, Wietzes P, Volkering F, Janssen DB (2000) Assessment of the use of partitioning and interfacial tracers to determine the content and mass removal rates of nonaqueous phase liquids. ES&T 34:4301-4306

O'Hara S, Krug T, Major D, Quinn J, Geiger CL, Clausen C, Yoon S (2004) Performance evaluation of dehalogenation of DNAPLs using emulsified zero-valent iron. Session C3, Remed. of Chlorin. and Recalcitrant Compounds, 4th Internat. Conf. Monterey, CA

Pankow JF, Cherry JA (1996) Dense chlorinated solvents and other DNAPLs in groundwater. Waterloo Press, Portland, Oregon

Parker BL, Cherry JA, Chapman SW, Guilbeault (2003) Review and analysis of chlorinates solvent dense nonaqueous phase liquid distributions in five sandy aquifers. Vadose Zone Journal 2:116-137

Payne FC, Suthersan SS, Lenzo FZ, Burdick JS (2001) Mobilization of sorbed-phase chlorinated alkenes in enhanced reductive dechlorination. Proceed. 6th International In Situ and On-Site Bioremediation Symposium 53-60

Poulson MM, Kueper BH (1992) A field experiment to study the behavior of tetrachloroethylene in unsaturated porous media. ES&T 26:889-895

Poulson SR, Naraoka H (2002) Carbon isotope fractionation during permanganate oxidation of chlorinated ethylenes (CDCE, TCE, PCE). ES&T 36:3270-3274

Quinn J, O'Hara S, Krug T, Geiger CL, Clausen C (2004) Evaluating the distribution of emulsified zero-valent iron for four different injection techniques. Session E4, Remed. of Chlorin. and Recalcitrant Compounds, 4th Internat. Conf. Monterey, CA

Ramsburg CA, Pennell KD (2002) Density-modified displacement for DNAPL source zone remediation: Density conversion and recovery in heterogeneous aquifer cells. ES&T 36:3176-3178

Ramsburg CA, Pennell KD, Kibbey TCG, Hayes KF (2003) Use of the surfactant-stabilized emulsion to deliver 1-Butanol for density-modified displacement of Trichloroethene. ES&T 37:4246-4253

Richardson RE, James CA, Bhupathiraju VK, Alvarez-Cohen L (2002) Microbial activity in soils following steam treatment. Biodegradation 13:185-295

Sabatini DA, Knox RC, Harwell JH, Wu B (2000) Integrated design of surfactant enhanced DNAPL remediation: efficient supersolubilization and gradient systems. JCH 45:99-121

Sale T, McWhorter DB (2001) Steady state mass transfer from single-component dense non-aqueous phase liquids in uniform flow fields. Water Resour. Res. 37:393-404

Schnarr M, Truax C, Farquhar G, Hood E, Gonullu T, Stickney B (1998) Laboratory and controlled field experiments using potassium permanganate to remediate trichloroethylene and perchloroethylene DNAPLs in porous media. JCH 29:205-224

Schwille F (1988) Dense chlorinated solvents in porous and fractured media. Lewis Pubslisher, Chelsea

Seol Y, Zhang H, Schwartz FW (2003) A review on in-situ chemical oxidation and heterogeneity. Environmental and Engineering Geoscience 6:39-51

Shook GM, Pope GA, Kostarelos K (1998) Prediction and minimization of vertical migration of DNAPLS using surfactant enhanced aquifer remediation at netraul buoyancy. JCH 34:363-382

Sorenson KS (2003) Enhanced bioremediation for treatment of chlorinated solvent residual source areas. In: Henry SM, Warner SD (ed) Chlorinated solvent and DNAPL remediation: Innovative strategies for cleanup, vol 837. ACS Symposium Series, 119-131

Stroo HF, Unger M, Ward CH, Kavanaugh MC, Vogel C, Leeson A, Smith BP (2003) Remediating chlorinated solvent source zones. ES&T 37:224A-230A

Taylor TP, Pennell KD, Abriola LM, Dane JH (2001) Surfactant enhanced recovery of tetrachloroethylene from a porous medium containing low permeability lenses 1. Experimental studies. JCH 48:325-350

Taylor TP, Rathfelder KM, Pennell KD, Abriola LM (2004) Effects of ethanol addition on micellar solubilization and plume migration during surfactant enhanced recovery of tetrachloroethene. JCH 69:73-99

Temples TJ, Waddell MG, Domoracki WJ, Eyer J (2001) Noninvasive determination of the location and distribution of DNAPL using advanced seismic reflection techniques. Ground Water 39:465-474

Tick GR, Lourenso F, Wood AL, Brusseau M (2003) Pilot-scale demonstration of cyclodextrin as a solubility-enhancement agent for remediation of a tetrachloroethene-contaminated aquifer. ES&T 37:5829-5834

Tuck DM, Iversen GM, Pirkle WA, Rulison C (1998) Time-dependent interfacial porperty effects on DNAPL flow and distribution. In: Wickramanayake GB, Hinchee RE (ed) Nonaqueous Phase Liquids Remediation of Chlorinated and Recalcitrant Compounds. Battelle Press, Columbus, OH, pp 73-78

U.S.EPA (1992) Estimating potential for occurence of DNAPL at Superfund Sites. (9355.4-07FS OP)

U.S.EPA (1998) In situ remediation technology: In situ chemical oxidation. 542-R-98-008

U.S.EPA (1999) Use of Monitored Natural Attenuation at Superfund, RCRA Corrective Action, and Underground Storage Tank Sites. OSWER-Directive 9200.4-17-P

U.S.EPA (2003) The DNAPL remediation challenge: Is there a case for source depletion? EPA/600/R-03/143

U.S.EPA (2004a) In situ thermal treatment of chlorinated solvents - Fundamentals and field applications. 542-R-04-010

U.S.EPA (2004b) Site characterization technologies for DNAPL investigation. 542-R-04-017

U.S.EPA (2004c) Treatment technologies for site cleanup: Annual Status Report, 11th edition. 542-R-03-009

U.S.EPA (2004d) DNAPL Remediation: Selected projects approaching regulatory closure. EPA 542-R-04-016

Vera SM, Werth CJ, Sandford RA (2001) Evaluation of different polymeric organic materials for creating conditions that favor reductive processes in groundwater. Biorem. J. 5:169-181

Wadley SLS, Gillham RW, Gui L (accepted) Remediation of DNAPL source zones with grannular iron: Laboratory and field tests. Ground Water

Wiedemeier TH, Haas P (2002) Designing monitoring programs to effectively evaluate the performance of natural attenuation. GWMR 22:124-135

Wilson RD, Mackay DM (1995) Direct detection of residual nonaqueous phase liquid in the saturated zone using SF6 as a partitioning tracer. ES&T 29: 1255-1258

Yan L, Thompson KE, Valsaraj KT, Reible DD (2003) In-situ control of DNAPL density using polyaphrons. ES&T 37:4487-4493

Yang Y, McCarty PL (2000) Biologically enhanced dissolution of tetrachloroethene DNAPL. ES&T 34:2979-2984

Yang Y, McCarty PL (2002) Comparison between donor substrates for biologically enhanced tetrachloroethene DNAPL dissolution. ES&T 36:3400-3404

Zhang W-X (2003) Nanoscale iron particles for environmental remediation: an overview. Journal of Nanoparticle Research 5:323-332

Zheng D, Carr CS, Hughes JB (2001) Influence of hydraulic retention time on extent of PCE dechorination and preliminary characterization of the enrichment culture. Biorem. J. 5:159-168

Printing: Krips bv, Meppel
Binding: Stürtz, Würzburg